KU-499-999
BELOW FLEET ST
SALISBURY SQ
OUSE
BUCKINGHAM
COCKSPURST
BREN
CKSPURST LONDON
COCKSPURST
PALL MALL
BATTERSEA
CHARING CROSS
LONDON CENTRAL AREA NEW ZEA
CRAYFORD
TRAFALGAR SQ
LEADENHALL ST
STRAND

HOMO BRITANNICUS

HOMO BRITANNICUS

The Incredible Story of Human Life in Britain

CHRIS STRINGER

ALLEN LANE
An imprint of PENGUIN BOOKS

To Katy, Paul and Tom, the latest generation of *Homo britannicus*

ALLEN LANE

Published by the Penguin Group
Penguin Books Ltd, 80 Strand, London WC2R ORL, England
Penguin Group (USA) Inc., 375 Hudson Street, New York, New York 10014, USA
Penguin Group (Canada), 90 Eglinton Avenue East, Suite 700, Toronto, Ontario, Canada M4P 2Y3
(a division of Pearson Penguin Canada Inc.)
Penguin Ireland, 25 St Stephen's Green, Dublin 2, Ireland (a division of Penguin Books Ltd)
Penguin Group (Australia), 250 Camberwell Road, Camberwell, Victoria 3124, Australia
(a division of Pearson Australia Group Pty Ltd)
Penguin Books India Pvt Ltd, 11 Community Centre, Panchsheel Park, New Delhi – 110 017, India
Penguin Group (NZ), cnr Airborne and Rosedale Roads, Albany, Auckland 1310, New Zealand
(a division of Pearson New Zealand Ltd)
Penguin Books (South Africa) (Pty) Ltd, 24 Sturdee Avenue, Rosebank, Johannesburg 2196, South Africa

Penguin Books Ltd, Registered Offices: 80 Strand, London WC2R ORL, England

www.penguin.com

First published 2006
1

Set in Minion and Trade Gothic
Designed and typeset by Smith & Gilmour, London
Original photography: Benoît Audureau
Additional picture research: Sarah Hopper
Colour Reproduction by Media Development & Printing Ltd
Printed in Great Britain by Butler & Tanner Ltd, Frome

A CIP catalogue record for this book is available from the British Library

ISBN-13: 978-0-713-99795-8
ISBN-10: 0-713-99795-8

CONTENTS

It had been a long, dry summer and the huge river had shrunk to a single deep channel, its many meanders now forming isolated water holes and reed beds. The land around the river estuary sloped gently down to a coast as it curved eastwards in a sweeping bay for a hundred miles, the sun shimmering brightly on the calm sea. A solitary man, naked and powerfully built, perched amongst the reddening leaves of a maple tree. He glanced briefly up to the sky, looking for the clouds that might at last promise that autumn rains were on their way to cool the parched landscape. But his most intense gaze was on the grassy riverbank below, where the women and children were hacking with small stone tools at the carcass of a baby hippo. They had smashed the skull open and were pulling out the brain, which was still warm and moist to their fingers and lips. His brothers and the older children had driven the hyaenas away, and the rest of the hippos, including the mother of this baby, had moved into deeper water. The man had eaten his fill already, but he kept his guard up – there were lions and wolves not far away, not to mention elephants, to judge from the sounds of splintering wood coming from the trees behind him.

This is not some scene from our ancestral African homelands of two million years ago, nor even from southern Europe a million years ago, although there are similarities in the animals and the climate. This is Suffolk, about 700,000 years before the recent emergence of evidence from sea cliffs near Lowestoft provided such a vivid picture of the landscape inhabited by the first Britons.

PROLOGUE
THE MYSTERY OF THE FIRST BRITONS

The ancient history of Britain seems to be all around us today. Newspapers, magazines, television documentaries and websites vie for our attention with stories of Vikings, Saxons, Romans and older tribes, stretching back to the Neanderthals of the last Ice Age. But away from homes and high streets, secrets of Britain's ancient history still lie waiting to be prised from under our feet, as we walk across roads, along cliff paths, on beaches and in quarries and caves. This book is about an entirely new prehistory of Britain being pieced together from the hidden evidence teased from layers of soil and rock accumulated through many millennia.

Homo britannicus presents the work of the Ancient Human Occupation of Britain project (AHOB). This project has brought together thirty archaeologists, palaeontologists and geologists at Institutes across the country in a unique collaborative network to reconstruct the most detailed calendar of human presence and absence in Britain yet achieved, using the latest techniques of scientific investigation. Although human fossils are very rare in Britain, evidence of human occupation is scattered over the landscape, preserved in ancient river deposits, and stored up in caves, in the form of stone tools and animal bones. Fossil remains can tell us what people looked like, stone tools can reveal details of their behaviour and adaptations, while associated sediments and animal remains can be analysed to unlock the secrets of ancient climates and environments. During the course of the project, many of the mysteries and uncertainties in our early history have been resolved. By showing the vulnerability of humans to past climate and environmental changes, AHOB is also providing a warning for the future, for the human race will face challenges every bit as serious in the near future.

Britain seems a verdant and timeless land for its inhabitants today, and we know from artefacts and fossils that our Stone Age predecessors lived here many hundreds of thousands of years ago. We might think, therefore, that the roots of the British people lie deep in British soil. Astonishingly, we now know they can be traced back less than 12,000 years, roots far more shallow than those of our neighbours on the Continent. The native peoples of America, Australia and Japan can all claim to have lived longer in their own lands than the British in theirs. Before 11,500 years ago, Britain was subject to some of the most rapid and violent swings in climate and environment in the entire history of the Earth. So vicious and challenging were these changes that time and time again people could only ever establish a temporary foothold here before being completely swept away, and Britain had to be recolonized about every 100,000 years. We have evidence that between 500,000 and 12,000 years ago humans were only here at all for about 20 per cent of the time; between 180,000 and 70,000 years ago, Britain was abandoned, completely empty of people for over 100,000 years.

For most of the last one million years, Britain was not an island. There would have been a wide land bridge to the European continent, so we cannot study the British record in isolation. The plants, animals and humans that we find in Britain during the ice ages, and the warm periods between, came via western Europe. The rich ice age records of neighbouring countries can thus provide us with valuable data on the sources for migrating populations, and the refuges to which early British inhabitants might have fled during the bad times.

AHOB is investigating a series of key questions about the first Britons, and this book looks at those questions, who is addressing them (see the Appendix), and how they are being answered. We have remarkable new evidence from East Anglia showing that humans arrived here earlier than anyone would have believed even a few years ago and lived in an environment with a balmy climate like that of southern Europe. This will be the subject of Chapter 1. An ancient buried landscape in East Anglia is being re-exposed and reconstructed to reveal a human presence that goes back a staggering 700,000 years – the oldest definite evidence of humans in northern Europe. These early inhabitants belonged to a primitive species whose only tools were shaped stones. At this time Britain was a peninsula of western Europe, when the now-vanished Bytham River flowed eastwards across East Anglia to a vast delta, draining into a northern sea where its waters joined those of the ancient Rhine.

The rich site of Boxgrove, in what is now Sussex, shows us that 500,000 years ago big game hunters also lived near the coast in a climate similar to the present day. Their story is told in Chapter 2. These strongly built people belonged to the species *Homo heidelbergensis* and hunted and used handaxes to butcher horse, deer and rhino. But their time came to an abrupt end when Britain soon afterwards suffered its worst ice age. About 450,000 years ago, a huge ice cap spread from the north, extinguishing the mighty Bytham River and pushing the proto-Thames southwards to its present course. Every human inhabitant must have fled southwards or perished, and it was not until about 400,000 years ago that a warm climate, and people, returned. In the intervening time, evolution had changed the peoples of Europe, and the new arrivals are recognizable as ancestors of the later Neanderthals. Some of them were still making handaxes, and on the banks of the nascent River Thames, at Swanscombe in north Kent, over 100,000 of these stone tools have been found. The climate of Britain was warmer than at the present day, with a very distinct fauna and flora, and life must have been good for the hunter-gatherers of the time – the subject of Chapter 3. There is evidence from stone tools that suggests more than one group of people may have entered Britain at this time, and comparing the fossil and archaeological records of France and Germany shows that there was considerable diversity in humans and their behaviour about 400,000 years ago.

The descendants of the Swanscombe people suffered the same fate as their predecessors when an ice age returned about 380,000 years ago. Again they came back when the climate allowed it, about 320,000 and 240,000 years ago. But for some reason these early Neanderthal people were now struggling to maintain their foothold in Britain, and, compared with the golden times represented by Boxgrove and Swanscombe, life seems to have been harder. One of the major focuses of AHOB research is this neglected and little-understood period of prehistory, since evidence is sparse compared with earlier and later periods. We do not yet know what lay behind the serious population crashes of this time, but we can speculate. The population crisis after 200,000 years ago was uniquely severe. Another ice age peaked about 140,000 years ago, but when the climate warmed and the big game returned, there were no people – the only time Britain was empty of humans in an interglacial period in half a million years. Deer, rhinoceros, elephants, hyaenas and even hippos were here, but no one was hunting them – not a single stone tool or butchered bone has been found from a period lasting over 100,000 years. This is perhaps the biggest puzzle that AHOB is trying to solve. We think Britain was an island for some of this period, perhaps even for the first time in its history, but can that alone explain the whole extent of human absence? What kept the Neanderthals out of Britain when they were apparently well established in neighbouring regions such as France and Belgium? Human decline and absence are the focus of Chapter 4.

Whatever the reason for their earlier absence, the Neanderthals did finally make it back to Britain. We know from the evidence of an extraordinary site in Norfolk that they were eating mammoth, reindeer and horse here 60,000 years ago. The rich record from Lynford even includes the remains of thousands of beetles that lived off heaps of mammoth dung, in a climate of mild summers but arctic winters, and these insects provide a vivid picture of the environment and climate of this time. But as the fluctuating climate moved towards a peak of severity 20,000 years ago, new kinds of people with different kinds of stone tools entered Europe and eventually Britain – the modern-looking Cro-Magnons. They were *Homo sapiens*, like us in appearance and behaviour and part of a global dispersal of our species that is being tracked in increasing detail through fossils, artefacts and genes in every inhabited continent. The Neanderthals may have clung on here until about 30,000 years ago but, soon after that, they disappeared for ever. New evidence is emerging across Europe of a time of dynamic change and even interaction between these distinct human species. With the wealth of new archaeological, environmental, dating and even genetic evidence, we are finally close to solving the mystery of the disappearance of the Neanderthals and the part we played in it. These topics are the subject of Chapter 5.

Our species *Homo sapiens* is known for its resourcefulness and adaptability, and yet the British Cro-Magnons were no more capable than their predecessors of surviving the worst of a British ice age. They vanished 25,000 years ago, as an ice cap a mile thick built up in Scotland, but returned as the climate briefly improved, about 15,000 years ago. At Creswell Crags in Derbyshire, the creativity of the Cro-Magnons is further revealed by cave art, the most northerly known in Europe, while in Somerset's Cheddar Gorge a darker picture emerges of cannibalism of both adults and children 14,000 years ago. These sites are the focus of Chapter 6. But the ice suddenly re-advanced one last time for a final cleansing of humans from Britain, some 13,000 years ago. Only after that, in the last 11,500 years, can we really talk about British ancestors living on British soil with real continuity. In Europe there is greater evidence of human persistence by the Cro-Magnons, and rich records of their culture in the form of campsites and cave art are found in France, Spain and Germany across to Russia. Archaeological and genetic evidence suggests, however, that even in mainland Europe the Cro-Magnons suffered change and crisis.

We have been truly fortunate to enjoy one of the most stable periods of climate during the last 11,000 years. This stability has allowed the development and spread of farming and urbanization across much of the world, and the accompanying explosive growth in human numbers. As we come to understand the sensitivity of the Earth's climate system, and particularly that of the North Atlantic, we can predict that the present stability will end very soon. The history of Britain and Europe over the last 700,000 years is littered with rapid and severe climatic changes, when apparently settled plant, animal and human communities were swept away in periods as short as ten years – much less than a single human lifespan. Some of the world's leading climatologists predict that we will see more climatic change in the next hundred years than we have seen in the previous 700,000. Experts are divided as to whether the whole Earth will experience a 'super-interglacial', warmer than anything over the last 50 million years, or whether developing changes in North Atlantic circulation will trigger a rapid and major cold event in the USA and western Europe, with much of the Atlantic freezing over in winter, as it did when the Cro-Magnons were last driven from Britain. This book will finish with an examination in Chapter 7 of how such changes would affect these regions and their human populations, showing that the effects will be every bit as severe as those that caused our predecessors to flee or die out. Chance events have been important in shaping our evolution, but we are now gambling with the future of the whole planet.

Let us begin our quest for *Homo britannicus* (the first Britons) with an introduction looking at how our knowledge of the prehistory of Britain has developed, and how the search for our earliest ancestors has involved many twists, turns and dead ends over the last two hundred years.

INTRODUCTION
IN THE BEGINNING

PRECEDING PAGES: *The Deluge* by Francis Danby, first exhibited in 1840. The Biblical Flood dominated attempts to explain prehistory and geology in the western world.

Our understanding of ancient history evolved in fits and starts, with faltering progress until a major watershed was reached in 1859. In that year a series of key papers was presented that finally established the antiquity of humans and the process of evolution. Those studies laid the groundwork of our present understanding of humanity's ancient beginnings. But 250 years ago, few were aware of, let alone interested in, the prehistoric landscapes waiting to be unveiled. Then, the Bible, and Greek and Roman sources, were seen as the ultimate authorities on ancient history.

'All that is really known of the ancient state of Britain is contained in a few pages. We can know no more than what the old writers have told us,' were the dismissive words of Dr Samuel Johnson in 1780, as reported by James Boswell. At that time, most people in the western world relied on the Bible as their primary source for ancient history, ranging from the Creation and the Garden of Eden, through the epic story of the Flood, to the dispersal and diversification of peoples following the collapse of the Tower of Babel. The idea of a catastrophic global flood seemed to explain the widespread sequences of rocks containing marine fossils and, through inundation, the remains of creatures that lived on land as well. But the Bible said nothing particular about the history of Europe and Britain. As the Danish Professor of Archaeology Rasmus Nyerup put it in 1806, 'Everything which has come down to us from heathendom is wrapped in a thick fog; it belongs to a space of time which we cannot measure. We know that it is older than Christendom, but whether by a couple of years or a couple of centuries, or even by more than a millennium, we can do no more than guess.' Using imaginative extrapolations from the Old Testament, ancient stone monuments in Britain were sometimes linked via the Celts to Japhet (one of the sons of Noah), while from the New Testament there emerged the legend that the young Jesus visited Glastonbury in Somerset with Joseph of Arimathea. Otherwise, for their local ancient histories, European antiquarians turned to classical writers such as Caesar and Tacitus, who wrote of ancient Britons, Gauls, Celts and Druids. The Druids in particular became associated with standing stones such as those at Avebury and Stonehenge in Wiltshire, and this is still with us today in the ceremonies of the Bards at the Welsh Eisteddfods, and the solstice festivals at Stonehenge, which are recent re-creations of lost histories.

The first step towards archaeology was the recognition that shaped stones, found when digging ditches or foundations, or ploughing fields, were not 'ceraunia' (thunderbolts) or elf shot but were actually tools shaped by ancient humans. Around 1590 Michele Mercati (Director of the Vatican Botanic Gardens) illustrated ceraunia, which we now recognize as stone tools ranging in age from 5,000 to 30,000 years old. But he broke with convention in saying that while most men believed that ceraunia were produced by lightning, he considered they had been 'broken from very hard flints in the days before iron was used for the follies of war'. Mercati was largely ignored, but sixty-five years later the Frenchman Isaac de la Peyrere was less fortunate when he wrote that ceraunia were the work of a pre-Adamite race of man – his books were publicly burnt in Paris in 1655. Around the same time, in Britain, Sir William Dugdale argued in his *History of Warwickshire* that such stones were 'weapons used by the Britons before the art of making arms of brass or iron was known'. This idea that there could have been a Stone Age before metal was discovered or used was reinforced as collections of artefacts brought back to European museums from colonies during the ages of exploration and enlightenment showed that there were peoples still alive who used stone rather than metals to make tools.

Possibly the first British description of a specific ancient stone artefact was of a handaxe. In 1715, John Bagford reported that an antique dealer, Mr Conyers, had found a 'British weapon' and the tooth of an elephant near what is now Gray's Inn Road in London. Bagford imagined that it had been the spear point of an ancient Briton who had used it to kill one of the Roman emperor

BELOW: Two of the handaxes from Hoxne described and illustrated by John Frere in 1797.

Claudius's elephants. At the end of the same century John Frere of Norfolk sent some flint implements that he had found just across the Suffolk border at Hoxne to the Society of Antiquaries in London. In an accompanying letter he said, 'if not particularly objects of curiosity in themselves, [these] must . . . be considered in that light from the situation in which they were found.' Frere observed that they had been found in undisturbed deposits twelve feet deep with strange animal bones, and presciently reported that 'the situation in which these weapons were found may tempt us to refer them to a very remote period indeed; even beyond that of the present world'. We can now recognize these as 400,000-year-old flint handaxes, but most of Frere's contemporaries in Britain were still content with biblical timescales for prehistory. Sir Thomas Browne spoke for the majority when he wrote in 1635, 'Time we may comprehend, 'tis but five days elder than ourselves, and hath the same horoscope with the world.' In 1650, using Old Testament genealogies, Archbishop Ussher calculated a date for the creation of the Earth and Man that received wide currency and was even inserted in the margins of subsequent editions of the Bible: 4004 BC. This meant that Man had then been in existence for less than six thousand years, and similar calculations showed that the post-Flood world had been in existence for only some four thousand years.

By about 1820, a more complex model was developing that suggested there had been successive 'creations' of living things, each one destroyed by a separate Flood. During the European 'Age of Enlightenment', between about 1750 and 1820, a wealth of knowledge about the natural world not only began to accumulate in private collections and public museums, but was also investigated systematically for the first time. The seeds were sown for the revolutions in geology and biology that were to follow a few decades later. Rather than rely on the voices of biblical and classical authority, antiquaries (the archaeologists of the day) began critically to compare the evidence they observed with the ancient sources of their education, and natural philosophers (the scientists of the day) began to look at life and the universe in completely new, and analytical, ways. Geology was still in its infancy but some scientists recognized that the biblical accounts might not be telling the whole story. A few tried to reconcile the developing fossil record with the different days of creation, while others developed a catastrophic explanation for the geological succession.

ABOVE: Georges Cuvier (1769–1832), 'Pope of Bones'.

In 1771, Father Johann Esper discovered human bones underlying those of extinct animals in the Gailenreuth Cave near Bayreuth in Germany. In a 1774 publication he asked himself, 'Did they belong to a Druid or to an Ante-diluvian or to a Mortal Man of more recent times? I dare not presume without any sufficient reason these human members to be of the same age as the other animal petrifactions. They must have got there by chance together with them.' The most influential French geologist and naturalist of the time, Baron Georges Cuvier (nicknamed 'The Pope of Bones'), argued that the human bones had indeed become mixed with the more ancient fossil ones. Even easier for Cuvier to dismiss had been the 1731 claim by Professor Johann Scheuchzer of Zurich, who believed he had uncovered the 'bony skeleton of one of those infamous men whose sins brought upon the world the dire misfortune of the deluge'. Cuvier showed that the fossil '*Homo diluvii testis*' (human witness of the Flood) was in fact the remains of an extinct giant salamander! He believed there had not been just one flood but several, each destroying a world that God had created earlier. Since humans were only present during the last of these (the Noachian Flood), it was hardly surprising that the biblical account was incomplete. This also meant that human bones could not have been fossilized during the earlier geological events and Cuvier strongly opposed any attempts to prove otherwise.

As influential in Britain in the 1820s as Cuvier was in France, the Reverend William Buckland, Professor of Geology at Oxford University and later Dean of Westminster, brought the beginnings of a scientific approach to the study of fossil bones from British caves. In *Reliquiae Diluvianae* ('Relics of the Flood: observations on the organic remains contained in caves, fissures, and diluvial gravel, and on other geological phenomena, attesting the action of an universal deluge'), he argued that superficial deposits of sands and gravels were evidence

ABOVE: Cuvier giving one of his brilliant expositions on anatomy in Paris.

of the global Great Flood. Like Cuvier he favoured the idea that there could have been earlier creations and floods, but that human remains would not be found in flood deposits (diluvium) except, perhaps, near the location of the Garden of Eden, after humans had been created. A brilliant but eccentric man, Buckland carried out some of the first proper investigations of diluvial deposits in caves, and excavated one of the first fossil human skeletons known to science, although the evidence was subjected to his own idiosyncratic interpretation. He must have been an extraordinary sight, travelling the length and breadth of Britain for his geological and mineralogical studies, astride a horse and dressed in top hat and professorial gown.

Buckland's view that geology was not only consistent with the Book of Genesis, but would also prove its truth, was reinforced by the conclusions he drew from his excavations at Kirkdale Cave in Yorkshire. Quarrymen had exposed a cave entrance and were using 'cattle bones' from the cave to stabilize muddy tracks. Buckland, however, had heard that these bones were peculiar

ABOVE: William Buckland (1784–1856) dressed for fieldwork, with his famous blue collecting bag.

and very numerous; were they perhaps the remains of creatures drowned in one of the great deluges? He set about exploring the complex web of cave passages, sampling and excavating as he went, and everywhere he found diluvial deposits packed with fossil bones. In keeping with his expectations, there were no human bones, but he found the remains of over twenty species of mammal, some native to Britain, some very exotic, and some quite extinct: they included deer, elephant, hyaena, lion and hippopotamus. Big game from Africa is very familiar to us now from zoos and wildlife documentaries, but this was not the case in 1821, and their discovery in a cave in Yorkshire was considered quite extraordinary. How such a peculiar mixture of creatures had arrived there was a challenging question – whether caused by a flood or not. One obvious but naive answer to this assemblage of alien bones would have been to assume that they had been washed from afar by the swirling flood waters. Buckland, however, made a brave scientific leap and argued that the bones represented animals that were alive at the same time in or around the cave. When the catastrophic Flood arrived, their remains outside the cave had been swept away but, in its deepest recesses, they had survived destruction.

The modern science of taphonomy is the study of the processes that occur from the time an organism dies to its burial and potential fossilization: these processes can include the manner of death, decomposition, movement, burial, and chemical alterations. Buckland was one of the pioneers of this science, although the Russian scientist Ivan Efremov did not coin the term until some 120 years later. Buckland noted that many of the bones in Kirkdale Cave were broken into small fragments and apparently had been gnawed, and he also noted that one of the most common animals represented as a fossil was the hyaena. He wrote to a colleague about the behaviour of modern hyaenas: 'As we know they do not dislike putrid flesh, we may conceive they took home to their den fragments of those larger animals that died in the course of Nature, and

LEFT: Kirkdale Cave as Buckland knew it (*above*). He recognized that the behaviour of modern hyaenas could help to explain the bone accumulations at Kirkdale, as shown in this contemporary caricature (*below*).

which from their abundance in the Deluge gravel we know to have been the Antediluvian inhabitants of this country.'

Not only was Buckland an early practitioner of taphonomy, he was also one of the first to conduct proper taphonomic experiments. He started by feeding various parts of animals to a Cape hyaena that was brought to Oxford in a travelling circus, and observing what was regurgitated from one end of the animal or dropped out of the other. Not content with the limitations imposed by the circus owner, Buckland imported his own hyaena from Africa, first intending to kill it and dissect its stomach contents and skeleton. But he could not bring himself to do this and kept Billy, as he named it, as a pet for the next

OPPOSITE: Dining at the Bucklands' must have been an interesting experience...

twenty-five years until its death. Billy performed his taphonomic tasks admirably, disturbing dinner party guests by chomping whole guinea pigs under Buckland's sofa, and there was soon a fine collection of chewed, regurgitated and defecated bones to compare with those found in Kirkdale Cave.

Buckland described the results of his excavations and experiments to the Royal Society in 1822, and published *Reliquiae Diluvianae* a year later, both to much acclaim. His work was of such popular interest that in 1825 another cave full of fossil bones was opened to the public in Banwell, Somerset. Profound lines by the landowner George Law, the Bishop of Bath and Wells, were written on the wall at the entrance:

> O thou, who, trembling, viewest this cavern's gloom,
> Pause and reflect on thy eternal doom,
> Think what the punishment of sin will be
> In the abyss of eternal misery.
> Here let the scoffer of God's Holy Word
> Behold the traces of a deluged world,
> Here let him in Banwell Caves adore
> The Lord of Heaven. Then go and scoff no more.

Banwell Bone Cave and its wonderful accumulations of fossils have taken on a new significance now, and will be discussed again in Chapter 4. But very different writing was already on the wall for Buckland and his views in the form of Uniformitarianism, the principle of which had been articulated by James Hutton in 1785: 'No processes are to be employed that are not natural to the globe; no action to be admitted except those of which we know the principle.' Buckland was aware of its growing influence, and argued in *Reliquiae Diluvianae* that it was impossible to attribute accumulations of deposits and bones 'to the effect of ancient or modern rivers, or any other cause, that are now or appear ever to have been in action, since the retreat of the diluvian waters'. In other words, the processes observed in the modern world simply could not be applied back to the extraordinary conditions of the Flood and pre-Flood worlds.
As Buckland started to lose ground in the face of new discoveries and a new generation of investigators, the challenges of interpreting evidence to fit his

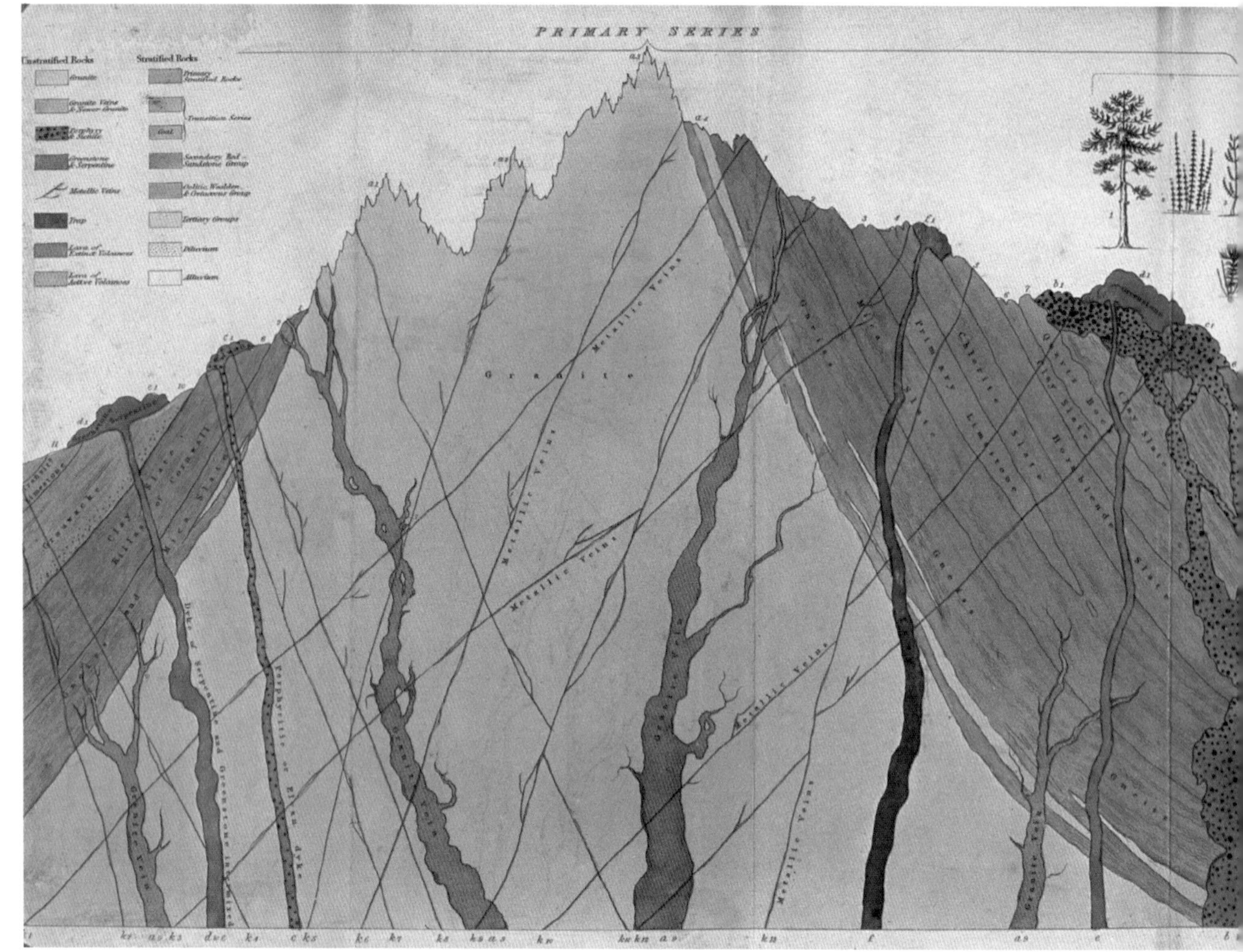

theories grew. In 1823 he even discovered a human skeleton at Goat's Hole, Paviland, in South Wales, in apparent association with mammoth ivory. We now know that this 'Red Lady' was in fact the 30,000-year-old burial of a Cro-Magnon man accompanied by red ochre pigments and ivory grave goods (see Chapter 5), but Buckland argued that it was the skeleton of a woman dating from Roman times, whose fellow tribesmen had dug up fossil ivory from the cave floor and fashioned it into jewellery: 'She was clearly not coeval with the antediluvian extinct species.'

Along with catastrophists like Cuvier and Buckland, there were other geologists who had gained their knowledge from engineering projects such as mining, quarrying and canal-building and who for entirely practical reasons

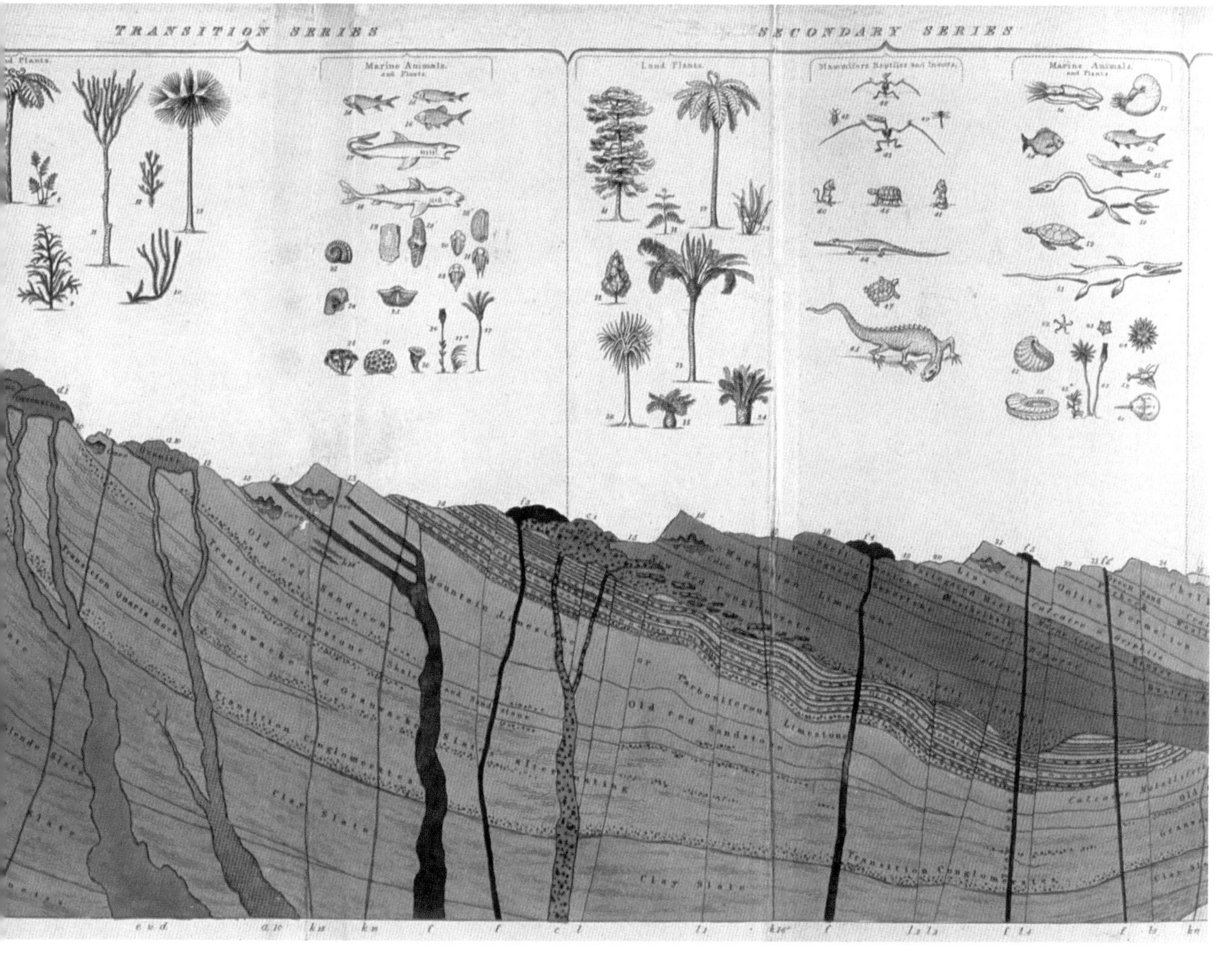

ABOVE: A geological diagram published by Buckland in 1836, showing the succession of rocks and fossils.

created classifications of the different rocks they encountered. In 1759 Giovanni Arduino (known as 'the father of Italian geology') wrote a letter in which he proposed a three-part division of the succession of rocks: Primitive or Primary; Secondary; and Tertiary. In 1799 the civil engineer William Smith (Arduino's English equivalent) began a classification of rocks that culminated in the publication of 'Strata Identified by Organized Fossils' in 1816, describing and mapping thirty-two superimposed layers of British rocks, each with their own index fossils. But the greatest revolution in geological thought came from a lawyer who was first inspired by Buckland's lectures to switch to geology and then increasingly began to challenge his teacher's views. As we shall see later, Charles Lyell's *Principles of Geology*, published in 1830, gathered a huge amount

ABOVE: The geologist Charles Lyell (1797–1875), portrayed in about 1835. His writings greatly influenced Charles Darwin, who was by then in the southern hemisphere on the *Beagle.*

of evidence in favour of Hutton's Uniformitarianism. He argued strongly that the present (not the Bible) is the key to the geological past.

A similar challenge to the primacy of biblical history was being mounted around the same time by archaeologists in Denmark who began to build up the framework of a succession of different technological ages that is still in use today. In 1813 the historian Vedel-Simonsen argued that 'the weapons and implements of the earliest inhabitants of Scandinavia were at first of stone and wood . . . later copper . . . and only latterly, it would appear, iron. Therefore from this point of view, the history of their civilisation can be divided into an age of stone, an age of copper and an age of iron.' Six years later Christian Thomson confirmed this division of different prehistoric periods when he arranged the collections of the Danish National Museum for display, and chose three successive Ages of prehistoric technology: Stone, Bronze and Iron. His views soon began to be influential in Britain, especially after an English translation of his Museum guidebook was published.

The Three Age system gave prehistory a depth and breadth well beyond that provided by the authority of the Bible or the accounts of Roman authors. When the Romans arrived in western Europe they had indeed encountered Gauls, Ancient Britons and Druids, but these were all peoples of the Iron Age, and there were at least two more ancient human worlds before theirs: the Bronze Age, and the even more ancient Stone Age. How ancient no one could yet say, of course.

One of the pioneers of British archaeology who began to flesh out the Stone Age and, tentatively, challenge Buckland's antediluvian views was a Catholic priest based at Torre Abbey in Devon. Father John MacEnery began digging in Kent's Cavern in Wellswood (now part of Torquay) in 1825, and soon broke through a floor of stalagmite to find stone tools associated with the bones of rhinoceros, sabre-tooth cats and other extinct animals. From his own accounts we can sense the struggle between excitement and scientific rigour as he examined some of the first finds: 'On tumbling it over, the lustre of enamel . . .

betrayed its contents. They were the first fossil teeth I had ever seen, and as I laid my hand on them, relics of extinct races and witnesses of an order of things which passed away with them, I shrank back involuntarily. Though not insensible to the excitement attending new discoveries, I am not ashamed to own that in the presence of these remains I felt more of awe than joy. But whatever may have been the impressions or the speculations that naturally rushed into my mind, this is not the place to indulge them – my present business is with facts.'

MacEnery was regularly in touch with Buckland about the progress of the excavations, and the Professor visited the cave on several occasions, even publishing some of the finds in support of his antediluvial theories. But when MacEnery wrote to him about publishing findings that appeared to demonstrate the contemporaneity of humans and antediluvian species, Buckland argued that ancient Celts must have dug oven pits in the stalagmite floor of the cave, and that some of their tools had worked their way down into the antediluvian deposits below. Although MacEnery continued the excavations under very difficult conditions for another ten years, he was so discouraged by Buckland's dismissal of the evidence that he abandoned publication of his findings: 'It is painful to dissent from so high an authority, and more particularly so from my concurrence generally in his views of the phenomena of these caves.' Fortunately his manuscript was saved and eventually published in 1859, when he finally got the recognition his insight and dedication deserved – but sadly that was already eighteen years after his early death, which was almost certainly accelerated by his years of work in the damp and dark depths of the cave.

Another West Country man, William Pengelly, took up the torch for British prehistory as it slipped from MacEnery's dying grasp. Pengelly was born in Cornwall but had relatives in Torquay, where he helped to found its Natural History Society in 1844.

BELOW: A signed portrait of William Pengelly (1812–1894).

BRIXHAM.

GREAT NATURAL CURIOSITY.

INTERESTING EXHIBITION ! !

THE

"Ossiferous Cavern"

Recently discovered on Windmill Rea Common, will be exhibited for a short time only, by Mr. PHILP, who has just disposed of it to a well-known scientific gentleman.

Those who delight in contemplating the mysterious and wonderful operations of nature, will not find their time, or money mis-spent, in exploring this remarkable Cavern, and as the fossils are about to be removed, persons desirous of seeing them had better apply early.

Many gentlemen of acknowledged scientific reputation, have affirmed that the stalactitic formations are of the most unique and interesting character, presenting the most fantastic and beautiful forms of crystallization, representing every variety of animal and vegetable structure.

Here too, may be seen the relics of animals that once roamed over the Earth before the post-tertiary period, or human epoch.

THE BONES AND TEETH, &c., OF

HYENAS, TIGERS, BEARS,

LARGE FOSSIL HORNS

of a Stag, all grouped and arranged by an eminent Geologist.

N.B. Strangers may obtain particulars of the locality, &c., of the Cavern, on application to Mr. BROWN, of the Bolton Hotel; or at the residence of the Proprietor, Spring Gardens.

THE CHARGE FOR ADMISSION TO THE "CAVERN," SIXPENCE.

Children will be admitted for FOURPENCE.

ated, Brixham, June 10th, 1858.

EDWARD FOX, PRINTER, &c., BRIXHAM.

ABOVE: One of John Philp's promotional posters for Brixham Cavern.

Two years later he would continue MacEnery's excavations at Kent's Cavern but his most important work began in 1858, work that was finally to extinguish Buckland's fading antediluvial theories. In that year, above the Devon fishing village of Brixham, a quarryman's crowbar fell down a crevice into a pristine cave. When the quarry owner John Philp enlarged the hole and squeezed down it, he found himself beneath a continuous ceiling of stalactite, crawling on ancient cave sediments. Cleaning the deposits under candlelight revealed hordes of fossilized animal bones, and he soon opened the 'Bone Cavern' at Brixham to the paying public. Pengelly realized the potential of a sealed and untouched site to resolve the question of the association of humans and antediluvial animals once and for all, and attempted to negotiate with Philp for permission to excavate the cave properly.

Very much a Del Boy of his time, Philp demanded the enormous sum (for those days) of £100, quite beyond the means of the Torquay Natural History Society. Fortunately, Pengelly had connections in high places and invited the distinguished geologist and palaeontologist Hugh Falconer to visit the site, after which Falconer was able to persuade the Geological Society of London and the Royal Society to sponsor the excavations. Meanwhile Philp printed and displayed posters all over Devon: 'The "Ossiferous Cavern" recently discovered on Windmill Rea Common will be exhibited for a short time only, by Mr. PHILP, who has just disposed of it to a well-known scientific gentleman. Those who delight in contemplating the mysterious and wonderful operations of nature, will not find their time, or money mis-spent, in exploring this remarkable Cavern, and as the fossils are about to be removed, persons desirous of seeing them had better apply early...'

With a curious public and the financial support of two of the most prestigious academic societies in Britain, Pengelly must have felt the weight of expectations as he began to excavate in the Brixham Bone Cavern. Within a few weeks, however, the team had collected not only a large number of fossil bones but also the first flint tools in association with them. Aware of his responsibilities, Pengelly developed a new recording system that he was to apply fully when he later returned to excavate in Kent's Cavern. His techniques required each specimen to be examined in place before removal, and its position accurately determined. When the usual approach was to pickaxe straight into the richest deposit and then bucket or wheelbarrow the finds away en masse, his methods were far ahead of their time, and his records are still used to reconstruct the stratigraphy and position of important finds from his excavations. In September 1858, Pengelly and Falconer's results were presented at a meeting of the British Association in Leeds, stating that they had found clear evidence of stone tools in association with the bones of hyaena, rhinoceros and other antediluvian species. Public and scientific interest was now primed for the even more remarkable events that were to unfold in the following year.

In Europe, events had been taking a parallel course, and Falconer and the geologist Lyell were keeping watchful eyes on developments there. Around 1830, work in French and Belgian caves had also turned up apparent associations of extinct species and artefacts. Pierre Tournal, working in the cave of Bize in the Midi region of France, not only found evidence of stone tools with the bones of cave bear, rhinoceros, hyaena and reindeer, but even carving on reindeer antlers. He noted that the finds showed signs of gradual and natural deposition in different layers, rather than by the action of a catastrophic flood. Near Liège in Belgium, Philippe-Charles Schmerling, a doctor of Austrian origin, worked in a number of caves beside the River Meuse. Like MacEnery and Pengelly, he found stone tools and fossils of extinct species such as rhinoceros and mammoth beneath unbroken stalagmite floors but, unlike them, he also found intermingled human remains in four of the caves. Making careful observation of the condition and preservation of the human bones, he was confident that 'the human bones were buried at the same time and by the same causes as the

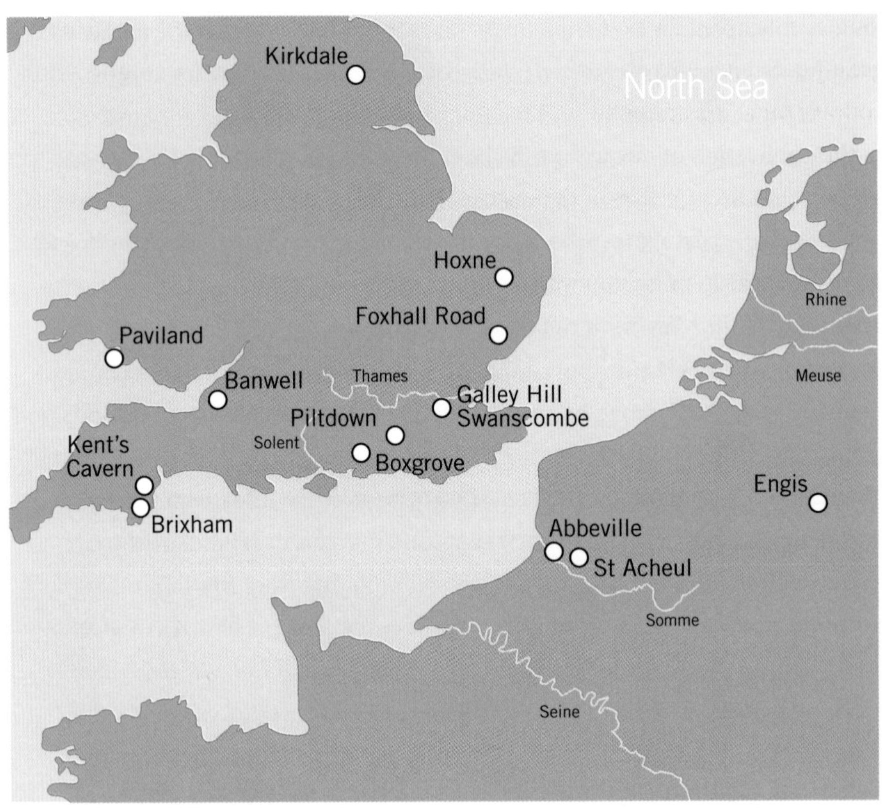

RIGHT: Britain and western Europe, showing some of the sites discussed in this chapter.

other extinct species'. Moreover, he added, 'Even if we had not found human bones in circumstances strongly supporting the assumption that they belonged to the antediluvian period, proof would have been furnished by the worked bones and shaped flints.'

Schmerling's most famous finds were made in the Engis Cave between 1829 and 1830, where he discovered two partial human skulls. Although direct radiocarbon dating now suggests that one is actually less than 10,000 years old, another (the skull of a child) was finally recognized in 1936 as belonging to a Neanderthal. The development of its teeth suggests that it was less than four years old at death, so the skull had not yet developed its characteristic adult shape, a shape that became known only after 1856. Charles Lyell visited Schmerling's excavations and cited them in the third edition of his *Principles of Geology* in 1834, but at that time he and most other scientists were still cautious about what the discoveries signified. Thirty years later, and sadly long after Schmerling's death, Lyell made amends and paid rich tribute to his great persistence and discoveries: 'at length, after finding leisure, strength, and courage for all these operations, to look forward, as the fruits of one's labour, to the publication of unwelcome intelligence, opposed to the prepossessions of the scientific as well as of the unscientific public . . . we need scarcely wonder, not only that a passing traveller [Lyell] failed to stop and scrutinise the evidence, but that a quarter of a century should have elapsed before even the neighbouring professors of the University of Liège came forth to vindicate the truthfulness of their indefatigable and clear-sighted countryman.'

It was the work of a Frenchman that finally pushed waverers like Lyell from their fence-sitting about whether humans were really contemporary with the fossils of extinct animals. Jacques Boucher de Perthes, a customs officer at Abbeville on the River Somme, had begun collecting handaxes from the river gravels in 1836, and ten years later he published *Antiquités celtiques et antédiluviennes*, in which he claimed that the stone tools were in association

AWFUL CHANGES.

MAN FOUND ONLY IN A FOSSIL STATE——REAPPEARANCE OF ICHTHYOSAURI.

A Lecture.—"You will at once perceive," continued PROFESSOR ICHTHYOSAURUS, "that the skull before us belonged to some of the lower order of animals; the teeth are very insignificant, the power of the jaws trifling, and altogether it seems wonderful how the creature could have procured food."

LEFT: A cartoon from the early years of palaeontology, putting a nice twist on the evolutionary story.

with antediluvian fossil bones. 'In spite of their imperfection, these rude stones prove the existence of man as surely as a whole Louvre would have done,' he argued. At first, his findings were questioned or ignored, but by the 1850s they had fomented and catalysed a fierce debate on both sides of the Channel. He received unexpected support in 1854 when one of his strongest and most distinguished opponents, Dr Jérôme Rigollot, announced that in searching for disproof he had instead made similar finds at St Acheul, near Amiens.

Then in 1858 Hugh Falconer took a winter trip to Sicily to improve his health. With Pengelly's discoveries fresh in his mind, he called in at Abbeville to see de Perthes' evidence, and was greatly impressed by the association of 'flint hatchets' and elephant molars. He invited Joseph Prestwich, a fellow member of the Geological Society of London, and the archaeologist John Evans to visit Abbeville and St Acheul with him in 1859. On his stomach-churning ferry voyage to France, Evans had written, 'Think of their finding flint axes and arrow-heads at Abbeville in conjunction with bones of Elephants and Rhinoceroses 40 ft below the surface in a bed of drift . . . I can hardly believe it. It will make my ancient Britons quite modern if Man is carried back in England to the days when Elephants, Rhinoceroses, Hippopotamuses and Tigers were also inhabitants of the country.' But Evans and Prestwich were completely won over by what they saw, and May and June 1859 provided a fever pitch of academic presentations at the Royal Society, the Royal Institution and the Society of Antiquaries in London. Falconer publicly supported the claims of de Perthes and Rigollot, Pengelly reported the results of his excavations at Brixham, and Prestwich presented a paper 'On the Occurrence of flint implements associated with the remains of animals of extinct species in beds of a late geological period at Amiens and Abbeville and in England at Hoxne'. Evans added a discussion of the stone tools and a few days later exhorted the people of Britain to find more evidence, saying, 'This much appears to be established beyond doubt, that in a period of antiquity remote beyond any of which we have hitherto found traces, this portion of the globe was peopled by Man.' At last Frere, Tournal, Schmerling and MacEnery were vindicated: whether under stalagmite floors in caves or deep in river or lake sediments, flint tools were clearly present in association with the bones of extinct (antediluvial) animals. Lyell was in the audience at the Royal Society meeting and finally followed suit publicly when he addressed the

Geological Section of the British Association in Aberdeen later in the same year, stating that he was 'fully prepared to corroborate the conclusions . . . recently laid before the Royal Society by Mr Prestwich'.

The year 1859 was thus critical for our understanding of human prehistory. Despite a few waverers and doubters, the tide finally turned in favour of the concept of humans as part of an ancient world inhabited by distinct and extinct faunas, and the floodgates were opening. The publication of Charles Darwin's book *On the Origin of Species* came in November. In his later autobiography Darwin confessed that he fully understood the implications of his ideas for human origins as far back as 1838, but that he was also aware in 1859 of the lack of supporting fossil evidence: 'Although in the *Origin of Species*, the derivation of any particular species is never discussed, yet I thought it best, in order that no honourable man should accuse me of concealing my views, to add that by the work in question "light would be thrown on the origin of man and his history". It would have been useless and injurious to the success of the book to have paraded without giving any evidence my conviction with respect to his origin.' Nevertheless, the ramifications of stating that all species were linked by common descent was there for all to see. Four years later, in 1863, Lyell showed the extent of his conversion to a long time scale for humans in *The Geological Evidence of the Antiquity of Man* and Thomas Huxley boldly went much further on human evolution than Darwin could have in 1859, in *Man's Place in Nature*. In 1865 Sir John Lubbock (later Lord Avebury) of the British Museum published *Prehistoric Times* and extended Christian Thomson's three ages of technology into four. He recognized that the Stone Age was of two kinds: an older period of flaked tools such as those found by Frere, Pengelly and Boucher de Perthes, and a later period of polished artefacts like those found in Danish burials, Swiss lake dwellings and the chambered tombs of Britain. He coined the terms Palaeolithic for the Old Stone Age, and Neolithic for the New Stone Age, both of which preceded the Bronze and Iron Ages. This chronology became widely accepted,

ABOVE: Charles Darwin (1809–1882), painted a year before his death.

SCOCIA : VLTRA MARINA
hec et albania dicta est
Sutherlande
Orkades Insule
Galewe
Regio scotorum
WALLIA
Karleola
Richemund
Pons Burgi
Pont fret
Ebora
Lincolnia
Lichefeld
Bristol
Bathonia
Hoiham
Dunestap
NORTH FOLK
SVFOLK
DE VONIA
DORSETE
CORNUBIA
london
CANCIA
Chanet
Vecta
ORIENS
AUSTER

A
DELINEATION
OF THE
STRATA
OF
ENGLAND AND WALES,
WITH PART OF
SCOTLAND;
EXHIBITING
THE COLLIERIES AND MINES,
THE MARSHES AND FEN LANDS ORIGINALLY OVERFLOWED BY THE SEA,
AND THE
VARIETIES OF SOIL
ACCORDING TO THE VARIATIONS IN THE SUBSTRATA,
ILLUSTRATED by the MOST DESCRIPTIVE NAMES
BY W. SMITH
THE GERMAN OCEAN
IRISH SEA
ST. GEORGE'S CHANNEL
CARDIGAN BAY
BRISTOL CHANNEL
THE ENGLISH CHANNEL
EXPLANATION

PRECEDING PAGES: Two historic maps of Britain. On the left is one of the oldest, prepared by the monk Matthew Paris in about 1259. On the right, 'The Map that Changed the World', according to author Simon Winchester: William Smith's 1815 map was the first geological map of any country.

but there were still many unanswered questions about the Stone Age. What kind of Britain could have been inhabited by such strange mixtures of creatures as hyaenas, reindeer and hippopotamus? How long ago was the Palaeolithic, and how long did it last? And what kind of people made the stone tools that Frere, MacEnery and Pengelly had excavated?

Answering those questions required a concept of recent Earth history that involved neither catastrophic floods nor a Britain that was just a slightly older version of that of the nineteenth century. In particular it had to explain how Britain could have been inhabited by animals that then were found only in the Arctic (reindeer) and the tropics (hippos). During the first half of the nineteenth century, just such a concept had in fact been emerging. And, in this case Dean Buckland can be seen as a progressive and enlightened figure rather than a religiously inspired opponent of new ideas. His view had been that the superficial deposits of the Earth (diluvium) had been produced by the Noachian Great Flood. But some geologists recognized that even on the tops of hills and mountains there were exotic rocks transported far from their place of origin. The idea grew that they might have floated to these places trapped in icebergs that had been displaced in the Flood. As the icebergs melted, they dropped their loads of 'erratics', and the material so deposited became known as drift. However, in the early 1800s some naturalists and engineers working in regions of Europe such as the Alps had noticed that drift deposits were still being produced where there were mountain glaciers.

In 1829 a Swiss civil engineer, Ignaz Venetz-Sitten, had recognized that landscapes well beyond the Alps seemed to show signs of glacial erosion and deposition, and in 1836 a Swiss amateur geologist, Jean de Charpentier, took a doubting young zoologist called Louis Agassiz to see this evidence. Agassiz, a former student of Cuvier, was soon converted from a diluvial to a glacial model for the creation of drift deposits. Buckland had invited him to Britain to study fossil fishes in 1834, and Agassiz now returned the compliment by inviting him to Switzerland to examine the glacial evidence. To his great credit, Buckland not only quickly acknowledged that Agassiz's interpretations were valid, he also said that he had seen similar formations in parts of Britain where there had been no ice in historic times. As they both realized, this meant that the Ice Age was not just a local phenomenon, and Buckland invited Agassiz back again, this time to visit Scotland and look at possible glacial landscapes. In 1840 Agassiz gave papers on both fossil fish and glaciers at the British Association for the Advancement of Science meeting in Glasgow and was soon ready to publish his observations showing that Scotland must once have been completely covered by ice. These observations included the sculpting of the landscape with valleys showing a characteristic U-shaped profile, rocks that had been scratched by passing glaciers, accumulations of huge numbers of erratics, and the presence of moraines (deposits of drift dumped in front of expanding ice caps).

Buckland's support certainly helped the acceptance of Agassiz's ideas in Britain, and he even converted his old student Charles Lyell to the idea. Nonetheless, both Buckland and Agassiz were still biblical creationists at heart, and neither believed in a great antiquity for humans, nor in Darwinian evolution. And Agassiz became so obsessed with the concept of an Ice Age that he saw its effects everywhere, even as far as the Amazon River. Equally, he was not able to abandon all of Cuvier's catastrophism. He wrote in 1866, 'The gigantic quadrupeds, the Mastodons, Elephants, Tigers, Lions, Hyenas, Bears, whose remains are found in Europe from its southern promontories to the northernmost limits of Siberia and Scandinavia . . . may indeed be said to have possessed the Earth in those days. But their reign was over. A sudden intense winter, that was also to last for ages, fell upon our globe; it spread over the very countries where these tropical animals had their homes, and so suddenly did it come upon them that they were embalmed beneath masses of snow and ice, without time even for the decay which follows death.'

Although Agassiz was able to convince important figures like Buckland and Lyell of the validity of his Glacial theory, it was evident to others that there had not been just a single worldwide Ice Age – things were clearly more complex than that. In 1854, the Swiss geologist Morlot suggested that there were two glacial stages separated by a warmer diluvial stage, and in Britain James Geike suggested in his 1874 book *The Great Ice Age and its Relation to the Antiquity of Man* that there had been a series of alternating glacial (cold) and interglacial (warm) stages. By 1909, two researchers Albrecht Penck and Eduard Bruckner proposed in their book *Die Alpen im Eiszeitater* (*The Alps in the Ice Age*) that there were four glacial stages in the Alps, which they named after Alpine rivers (in increasing age) Würm, Riss, Mindel and Günz, with intervening warmer

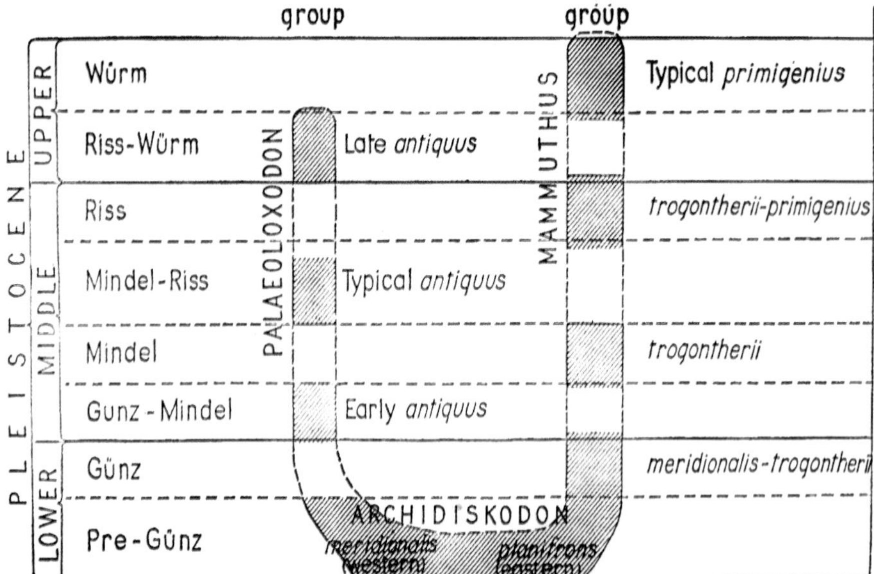

RIGHT: This diagram from 1964 used the simple Alpine sequence of glacials and interglacials to interpret the evolution of straight-tusked elephants and mammoths. (An AHOB chart showing geological subdivisions and archaeological stages can be found after the Appendix.)

stages called Riss-Würm, Mindel-Riss and Günz-Mindel, respectively. As we will see, this division liberated geology and archaeology from the straitjacket of Agassiz's single great Ice Age, but eventually the four-ice-age system became a straitjacket of its own, as scientists applied the Alpine model globally, despite growing evidence for even greater complexity in past climate change.

The term Quaternary ('Fourth') has become popular for the period of Earth history containing the glaciations and extending up to the present day, following on from the three stages Arduino had named in 1759 (Primary, Secondary, Tertiary). Lyell proposed the alternative term Pleistocene in 1839 from the Greek words for 'most' and 'recent'. Yet another term was added later to cover the period we live in today following the ice ages: the Holocene ('wholly recent'). Thus the Quaternary is equivalent to the Pleistocene + Holocene. In the rest of this book we will use Pleistocene to cover the period of time from about 1.8 million years ago to the end of the last Ice Age, about 11,500 years ago, at which point we come into the Recent or Holocene. The Pleistocene is also nearly synonymous with the Palaeolithic (Old Stone Age), but we know that early humans were actually using stone tools in Africa more than two million years ago, so the Palaeolithic extends back beyond the Pleistocene into the preceding Pliocene.

In 1867 Gabriel de Mortillet, a Professor of Anthropology in Paris, further developed the divisions of the Palaeolithic, based on layered deposits in caves and river gravels where different types of stone tools and associated animal remains could be sequenced in time in relation to each other. He argued that these represented epochs of increasing age in the Palaeolithic and named them after typical sites in France such as the Magdalenian (after the cave of La Madeleine = Reindeer Age); Mousterian (after Le Moustier Cave = early part of Cave Bear-Mammoth Age); Chellean (after the river deposits of Chelles = period of the straight-tusked elephant). Later these Stone Age periods were grouped into a new three-part classification of the Palaeolithic: the blade tools and art of the Magdalenian became part of the Upper Palaeolithic; the flake tools of the Mousterian became the Middle Palaeolithic; and the handaxes of the Chellean (similar to those found by Frere and Boucher de Perthes) became the Lower Palaeolithic. But fossils to show who made these different tools were remarkably slow in arriving, and in some cases they were positively misleading.

We now know that Buckland's finds from Paviland and Schmerling's from Engis include genuine Pleistocene fossils, but at the time they were either misinterpreted or ignored. Similarly, in 1848, a strange-looking skull was found after blasting in a limestone quarry below the sheer north face of the Rock of Gibraltar. But it was not properly examined for another fifteen years. In the meantime, some other peculiar bones were found in Germany that were eventually recognized for what they were – the remains of an ancient European people, the makers of Middle Palaeolithic tools. Near the town of Düsseldorf, the River Düssel runs through a small limestone valley that in the 1800s was peppered by cave openings. Named after the poet and hymn composer Joachim Neander, the Neanderthal (Neander Valley) began to be extensively quarried in 1854, and two years later, the clay filling one of the caves (Kleine Feldhofer Grotte – the small cave near the Feldhof Farm) was being thrown 60 foot down to the valley floor. Bones thought to be those of cave bears were seen as they were being dumped, and they found their way to a local teacher and natural historian, Johann Carl Fuhlrott, who recognized that they were actually human. In 1857 both he and an Anatomy Professor at the University of Bonn, Hermann Schaaffhausen, published descriptions of the Neanderthal human skullcap and partial skeleton. Schaaffhausen said that the bones 'exceed all the rest in those peculiarities of conformation which lead to the conclusion of their belonging to a barbarous and savage race', and he regarded them 'as the most ancient memorial of the early inhabitants of Europe'.

Others were less impressed with the bones. Professor F. Mayer of Bonn claimed the skeleton was probably that of a rickety Mongolian Cossack 'who, on his way through Germany towards France in 1814, had crept into the cave and died'. He came to this conclusion by observing the bowed leg bones, indicating horsemanship, an injured elbow (obviously a war wound), and the large brow ridges, evidence of the agony the individual had suffered before death as he frowned in pain! In contrast Lyell more soberly concluded, 'on the whole I think it probable that this fossil may be of about the same age as those found by Schmerling in the Liège caverns', while Thomas Huxley regarded it as representing no more than an extreme variant of *Homo sapiens*. William King, an Irish anatomist, considered that the shape of the skull was distinct enough to indicate a separate species, and at the 1863 meeting of the British Association for the Advancement of Science in Newcastle he proposed the name *Homo neanderthalensis*, the first new species of fossil human, a name that was published the following year. In doing so he gained priority for the Neander Valley finds and their name, since he pre-empted Falconer and George Busk, who were considering giving the name *Homo calpicus* (after the ancient name for Gibraltar) to the skull found there in 1848, which was actually that of a Neanderthal woman. As finds of similar-looking humans turned up in increasing numbers from caves in Belgium, France and Croatia, the 'Neanderthals' were gradually recognized as distinct and ancient inhabitants of Europe, although it was still unclear if even older people had existed anywhere else in the world.

ABOVE: Joachim Neumann (1650–1680) wrote the hymn *Praise to the Lord, the Almighty*. He preferred the classical version of his name – Neander – and the Neander Valley (Neander thal) was named after him.

By 1997 most of us thought that there was nothing new to add to the story of the Neander Valley finds; the site had been completely destroyed, no one had saved any associated materials such as stone tools, and the existing sixteen bones had been studied to death. But two things were radically to change the situation and put Fuhlrott's discoveries back in the centre of science. First they were sampled for ancient DNA, and as we shall see in Chapter 5 this gave us a completely new insight into the Neanderthals. Second, two German archaeologists had decided to attempt an incredible piece of detective work: they would try and locate the material lost from the cave in 1856. They had to examine the surviving quarry plans, look at sketches and paintings of the valley as it was before quarrying, and then drop deep trenches in the areas where the spoil might have been dumped from the cave down to the valley floor. One of them, Ralf Schmitz, wrote to me about the plans and put the chances of their success at about 5 per cent: incredibly, in one of their first trenches they found a piece of bone that fitted on to the original Neanderthal femur, where it was broken. This was like recovering one of the lost arms of the Venus de Milo. Soon they had excavated many stone tools and fossil bone fragments. These included the first human teeth and jawbone fragments from the site, and a piece of face that fitted perfectly on to the 1856 skullcap. Arm bone fragments duplicated those already found and this indicated that there was at least one more individual – demonstrated from the DNA recovered from it to be yet another Neanderthal. Three fragments were radiocarbon dated and gave similar dates of about 40,000 years old. Spanning 140 years, a connection of discovery was made between Fuhlrott and Schmitz, a physical connection was made between the old finds and the new, and the type specimen of the Neanderthal group once again became the centre of scientific attention as it yielded up the first ever DNA of an ancient human for analysis.

In 1868, twelve years after the discoveries in the Neander Valley, work on a railway cutting in the Vézère valley of the Dordogne region in France exposed a rock-shelter of the Cro-Magnon period full of the remains of reindeer, lion and mammoth, together with Upper Palaeolithic tools and ancient hearths. Amongst these finds were human bones, covered in red ochre pigment, and rows of seashells, apparently necklaces. However, in contrast to the Neanderthal finds, these bones looked much more like modern humans – and the site gave its name to these early Europeans. The remains were eventually published in detail in 1875 in a beautifully illustrated monograph *Reliquiae Aquitanicae*, but sadly only after the authors Edouard Lartet, a French magistrate turned palaeontologist, and Henry Christy, an English banker and hat-maker, had both died. The finds included engravings of reindeer and mammoth on fossil remains of the animals themselves, providing the most direct testimony of the contemporaneity of humans and ice age animals. Even more powerful testimony followed in 1880 with the publication by the Spanish prehistorian Marcelino Sanz de Sautuola of wonderful cave paintings of bison and bulls accidentally discovered by his young daughter in the Altamira Cave near Santander. Unfortunately, their sophistication meant that it was many more years before they were generally accepted as authentic.

Since, in many French sites, Middle Palaeolithic flakes and scrapers always seemed to underlie blade tools of the Upper Palaeolithic, it began to be argued that just as the Middle Palaeolithic gave way to the Upper Palaeolithic, so the Neanderthals must have given way to the Cro-Magnons. But what remained unknown was who had made the distinctive handaxes of the even more ancient Lower Palaeolithic. Would remains ever be found of the most ancient humans, ones who provided a link to our ape ancestry? In 1863 there had briefly seemed to be an answer to that mystery from the Moulin-Quignon quarry at Abbeville, the area of the Somme River studied by Rigollot and Boucher de Perthes. A human lower jaw had been found in apparent association with several handaxes, but while many French workers became convinced that the association was genuine, Evans, Prestwich and Falconer, previously so supportive of Boucher de Perthes, now demurred. They cautioned that the jaw looked like that of a modern human, and according to Busk and Falconer was 'gelatinous' like fresh bone, while the handaxes seemed to be recent forgeries, probably made using metal hammers and without the natural staining of genuine examples. Evans suggested that Boucher de Perthes put one of his most trusted workers on a careful watch and within a week there was clear evidence that handaxes were being planted. In 1863, Evans pronounced, 'I sincerely hope that the human jaw from Moulin-Quignon may from this time forward be consigned to oblivion. *Requiescat in pace!*' When chemical tests were applied to it in 1950, they confirmed that it was not fossilized.

There were claims from Britain of an even greater antiquity for another modern-looking jawbone, this time found in sands being excavated at Foxhall, near Ipswich. Again Falconer urged caution, and the high organic content of the bone seemed to warrant this. The story was repeated yet again when a human skeleton was discovered in a gravel pit at Galley Hill in Kent. As we shall see in

the following chapters, this part of Kent around the village of Swanscombe is famous for the thousands of handaxes it has produced, and in 1888 human remains were found in the same gravel deposits. They were robustly built but much more modern looking than the skeleton from the Neander Valley, and opinion was divided about whether this was a genuine fossil relic of the handaxe makers, or a much later burial cut into the river gravels. Chemical tests applied in the 1950s showed that the Galley Hill bones were not fossilized, and finally in 1960 a radiocarbon date was carried out on one of the arm bones, giving an age of only about 3,500 years. Unfortunately the current whereabouts of the Foxhall jaw is unknown, so it has not been possible to apply the same kinds of analyses to it.

Eventually, near the end of the nineteenth century, a really primitive human fossil *was* found, not in Europe but on the other side of the world. A Dutch army doctor, Eugene Dubois, used his posting to the Dutch East Indies (now

ABOVE: An ancient fossilized sand dune in Ayrshire: such deep and distinctive geological deposits constantly raised questions about the age of the Earth, and of humans.

Indonesia) in 1887 to search for the *Pithecanthropus* (ape-man) that the German biologist Ernst Haeckel had predicted would be found in southern Asia. His finds consisted of some teeth and jaw fragments, a human-looking thighbone and a skullcap even lower and flatter than the Neanderthal one, with huge brow ridges and a much smaller brain capacity. Dubois christened these finds *Pithecanthropus erectus*, now better known as Java Man or *Homo erectus*. So, much to the consternation of those European experts who recognized that the Neander Valley fossil was too human to represent a missing link, their continent had produced nothing to indicate it was important in human beginnings. So where *were* the remains of the people who had made the earliest stone tools, people who dated not from tens of millennia ago, but perhaps from hundreds or even thousands of millennia?

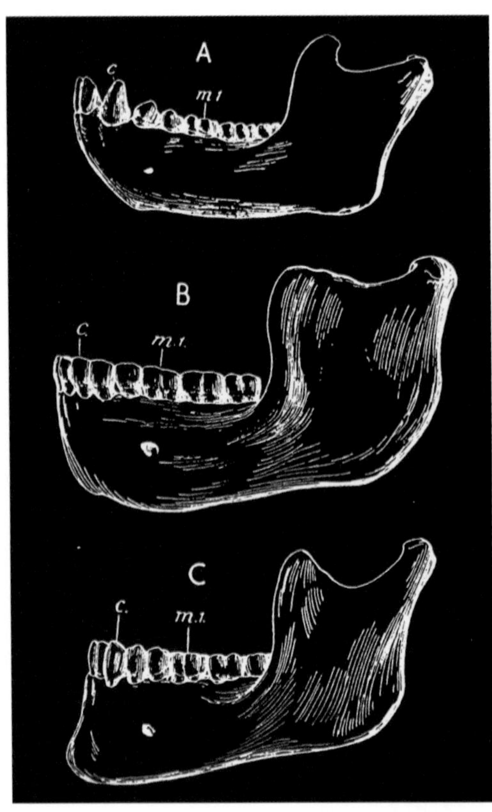

ABOVE: The massive jawbone from Mauer (Heidelberg) is compared here with the mandibles of a chimpanzee (*top*) and modern human (*bottom*).

In 1907, a possible answer emerged, and this time there was no question that this might be an intrusive modern human masquerading as an ancient fossil. Dr Otto Schoetensack, a geologist at Heidelberg University in Germany, had regularly examined animal bones found during quarrying in a sand pit at Mauer, a few miles from Heidelberg. At a depth of eighty feet, workmen discovered a very thick jawbone containing most of its teeth, and in contrast to the finds from Moulin-Quignon, Foxhall and Galley Hill, this jaw had absolutely no sign of a modern chin. Schoetensack studied the Heidelberg jaw, as it became known, and he published a thorough account of the discovery and its context in 1908, naming the jaw as the type of a new human species *Homo heidelbergensis*. The associated warm-climate animals included hippo, elephant, rhino, scimitar-tooth cat, red deer and two species of beaver. By 1908, there was a better understanding of the succession of mammal species in Europe, and Schoetensack argued that the Mauer fauna was much more ancient than ones known from the late Pleistocene such as those found with Neanderthal and Cro-Magnon fossils, and with Upper and Middle Palaeolithic tools. *Homo heidelbergensis* was from the Middle Pleistocene and by implication (since no definite artefacts were found at Mauer) was representative of the people who had made Lower Palaeolithic tools such as handaxes. Although Schoetensack did not attempt to estimate the actual age of the find, others guessed it could be ten times older than those of the Neanderthals and Cro-Magnons – perhaps an astonishing 500,000 rather than 50,000 years old. Some scientists recognized similarities to the Neanderthals, perhaps hints that this might be an earlier member of the same evolutionary line; others such as the British anatomist Sir Arthur Keith thought that the teeth indicated quite an advanced kind of human, while the Cambridge anthropologist Wynfrid Duckworth said, 'Would the Mauer jaw be appropriate to the cranium of *Pithecanthropus*? I believe an affirmative answer is justified.' Others, such as the Australian expert on the evolution of the brain Grafton Elliot Smith, believed that none of these finds offered evidence of the ancestor of our species *Homo sapiens*. Our large brain and distinctive globular skull form must have taken a long time to evolve, he argued, and somewhere in the fossil record from the beginning of the Pleistocene the evidence was waiting to be discovered.

There was an additional complication raised long before by people like Joseph Prestwich, whom we encountered in the ferment of the year 1859. Were the handaxes being found in the gravels of many European rivers really the oldest evidence of human occupation? For Prestwich, the answer was no. He believed that the most primitive humans would have made much more primitive stone tools, minimally modified cobbles and flakes. Such tools became known as eoliths ('dawn stones') and the search for them and their manufacturers preoccupied many archaeologists around the beginning of the twentieth century, particularly in Britain. James Reid Moir, President of the Prehistoric Society of East Anglia, scoured the cliffs and beaches of East Anglia for them, while Benjamin Harrison, a Kent grocer, searched the high 'plateau gravels' of south-east England. Eoliths were so simple that they would be almost indistinguishable from natural flakings, and this problem was not lost on the fiercest critics of their very existence. John Evans laid out the hallmarks of human workmanship that would be expected on genuine ancient tools – the physical signs left when stone was purposefully and directionally struck on stone by Man. While the signs were unquestionably there on classic artefacts from the

Lower, Middle and Upper Palaeolithic, they were sadly lacking on British eoliths. And in 1905, the French anthropologist Marcellin Boule not only suggested that apparent eoliths would be produced in large numbers when cobbles smashed against each other on riverbeds and seashores, he actually found perfect examples in cement-mixing mills at Nantes. As one French prehistorian remarked: 'Man made one. God made ten thousand. God help the man who can distinguish the one in the ten thousand.'

By 1912, British archaeologists, whether they favoured handaxes or eoliths, were desperate for a convincing early Briton. It was becoming a serious national embarrassment. Britain's greatest rival, Germany, had the Neanderthal skeleton and the Mauer mandible, our old imperial rivals the French had the Cro-Magnon finds and, from 1908, a succession of spectacular Neanderthal skeletons; even the Dutch had *Pithecanthropus* from the Dutch East Indies. Thus the stage was set for something that, in the end, would provide even more national embarrassment: Piltdown Man. This notorious affair was probably spawned by the finds of Java Man and Heidelberg Man, with the idea of creating an even more spectacular find on British soil. Charles Dawson, a solicitor and amateur fossil hunter, claimed that sometime before 1910 a workman handed him a thick, dark-stained piece of human skull that had been found in gravels

RIGHT: Charles Dawson (*left*) and Arthur Smith Woodward (*right*) excavating at Piltdown.

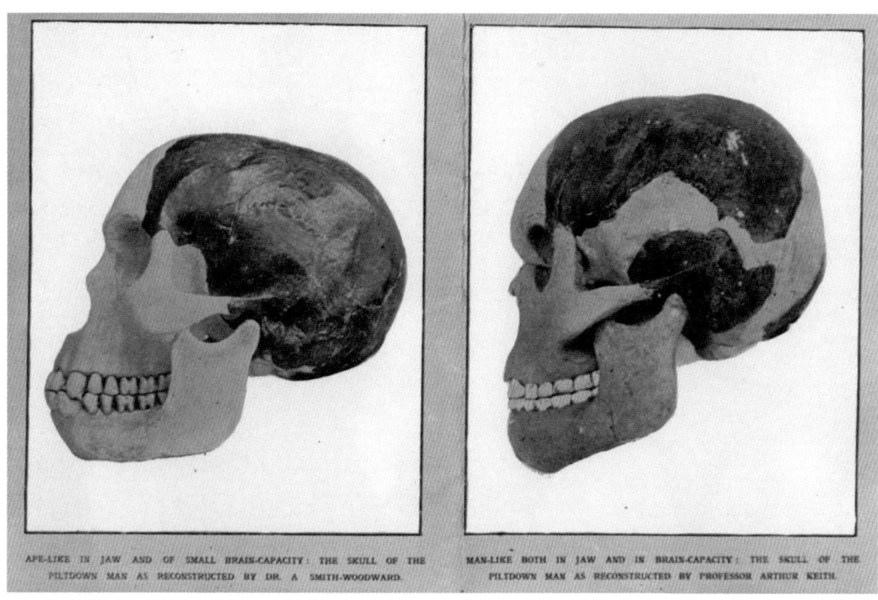

LEFT: Two early attempts at reconstructing the Piltdown skull by Smith Woodward (*left*) and Arthur Keith (*right*).

at the village of Piltdown in Sussex. By 1912, Dawson had collected more of the skull from around the site, and had contacted his friend Arthur Smith Woodward, Keeper of Geology at the British Museum (now the Natural History Museum, where I work). Together they began excavations at Piltdown in 1912, and soon found more skull fragments, fossil animal bones, stone tools, and a remarkable fragment of lower jaw.

Amid great excitement, they announced the finds to a packed session of the Geological Society in London at the end of 1912, and named a new type of early human, *Eoanthropus dawsoni* ('Dawson's Dawn Man'). Although the skull and jaw pieces were awkwardly broken, Smith Woodward reconstructed them into a complete skull that combined a rather modern-looking braincase with very ape-like jaws. On the basis of the associated animal bones and stone tools, Smith Woodward and Dawson argued that *Eoanthropus* dated from the early Pleistocene (some guessed as far back as a million years) and was thus more ancient than Heidelberg Man. Some experts remained doubtful, but in 1913 and 1914 more finds were made at Piltdown, including a canine tooth intermediate in size between that of apes and humans, and a unique carved artefact made from a large piece of elephant bone that because of its shape became known as the cricket bat. In 1915 the last Piltdown finds were made: a tooth and some skull pieces closely matching the first finds. These were

ABOVE: This painting shows a discussion about the Piltdown remains, with Arthur Keith at the centre, watched over by Elliot Smith (*to his right*) and Charles Dawson and Smith Woodward (*to his left*). All have been implicated in the hoax at one time or another.

supposedly found by Dawson in a field two miles from the original site. The additional finds swung the opinion of many sceptics in favour of Piltdown Man, and it helped that *Eoanthropus* met the expectations of scientists like Elliot Smith that the brain had evolved to a large size early in human evolution, while other features (such as the jaws and teeth) may have lagged behind.

But the days of *Eoanthropus* were numbered. As further finds of possible human ancestors were made in Africa and Asia during the 1920s and 1930s, Piltdown Man was pushed into an increasingly peripheral position in the story of human evolution, since nothing else resembled it. When new chemical and physical dating techniques were applied to Piltdown Man from 1949 onwards, the results were puzzling, suggesting that the skull and jaw material, unlike the fossil animal bones from the site, could not be very ancient. Then in 1953, their suspicions aroused, Oxford scientists Joe Weiner and Wilfrid Le Gros Clark asked Kenneth Oakley, Smith Woodward's successor at the British Museum, to apply even more stringent tests. They soon published their initial investigations and conclusions: the ape-like Piltdown mandible was a forgery. Over the next two years Oakley and colleagues conducted even more wide-ranging analyses which showed that the whole Piltdown assemblage of bones and artefacts was fraudulent. The site had been systematically salted with bones and artefacts from various sources, most of them artificially stained to match the colour of the local gravels. The 'missing link' itself consisted of parts of an unusually thick but quite recent human skull, and the jaw of an unusually small orang-utan with filed teeth. Uproar followed, and the press had a field day, with reactions ranging from mockery to questions in Parliament about the competence of the British Museum.

So who was responsible for this hoax, which for forty years fooled some of the most outstanding British scientists? At least twenty-five men have been accused of being involved in the forgery, ranging from Dawson and Smith Woodward through to the eminent anatomists Arthur Keith and Grafton Elliot Smith. Even Sir Arthur Conan Doyle, the creator of Sherlock Holmes, who lived in Sussex and played golf at Piltdown, has been added to the growing list of suspects. Dawson, however, remains the prime candidate for the forger. He was the first person seriously to search for and report fossils from the site, and he was present when all the main finds were made. He is the only individual who can definitely be associated with the final 'discoveries' at the second Piltdown site, and subsequent to his final illness and death, no further significant discoveries were made at either Piltdown locality. However, an alternative candidate has recently come to the fore in Martin Hinton, who at the time of the discoveries was a knowledgeable volunteer in Smith Woodward's department at the British Museum, and later became Keeper of Zoology there. In the mid-1970s an old canvas travelling trunk with his initials on it was found when loft space was being cleared above the old Keeper of Zoology's office. Amongst the items unpacked by AHOB member Andy Currant were mammal teeth and bones stained and carved in the manner of the Piltdown fossils.

The palaeontologist Brian Gardiner of King's College, London, has argued that the staining procedures in Hinton's materials were the same as those used in the Piltdown assemblage, and thus that Martin was the forger. His motive might

RIGHT: A Geologists' Association visit to Piltdown on 12 July 1913.

have been revenge following a quarrel about departmental payments due to him or perhaps he, like several others, had taken a personal dislike to Smith Woodward. Shortly after the exposure of the forgery, Hinton indicated in conversations, interviews and correspondence that he had long had suspicions about Piltdown and may even have known who was behind it. He certainly had the geological knowledge and access to materials to produce the forgeries, whether in league with Dawson or not. But there are now several additional reasons to suspect that Dawson was not merely the innocent victim of the malice or trickery of others. As mentioned already, he was the only figure present throughout the main events, and the strange 'discoveries' at Piltdown II can only be laid at his door. Additionally, there is now plenty of other evidence that Dawson was not the straightforward solicitor and honest amateur scientist he seemed. He was involved in a chain of actual or likely forgeries of fossils and archaeological objects, and in misrepresentation or plagiarism of the work of others. His motive for Piltdown was probably scientific and personal ambition, and knowing the field well he was able to create material that closely matched prevailing ideas and scientific agendas.

But what about Hinton? His attitude towards Piltdown both before and after the exposure was suspicious, and the contents of the trunk show he was experimenting with the faking of fossils – was this to create his own forgeries or to show how Piltdown could have been done? This brings us back to the extraordinary 'cricket bat', the last significant find at Piltdown I. Letters and interviews show that Hinton was well aware of the forgeries at Piltdown and, from the similarities with material in Hinton's trunk, I think he made and planted this absurd object to warn the forger(s) that the game was up. To his horror, instead of terminating the whole Piltdown saga, this bizarre piece was heralded as the world's oldest bone implement. Under this scenario, Piltdown II followed as Dawson's reaction to the contamination of the original site. He then fell ill and died before he could properly develop a new *Eoanthropus* somewhere else.

ABOVE: The Piltdown Man pub is still open in nearby Uckfield, East Sussex.

Of the original protagonists, only Sir Arthur Keith, Hinton and Teilhard de Chardin, a Jesuit priest and palaeontologist who had helped at the site and found the canine tooth, were still alive when the fake was exposed in 1953. Poor Keith lived to see both his favoured claimants for the earliest Britons – the putative eolith-maker from Piltdown, and the handaxe maker from Galley Hill – eliminated by new finds and new techniques. But he also lived to see and study the real thing, a discovery made a few hundred feet away from the spurious site of Galley Hill. There in the Barnfield Pit at Swanscombe, where thousands of handaxes had been found, parts of a skull were excavated in 1935 and 1936 that at last showed what the handaxe makers of Britain looked like.

This chapter has covered a lot of ground. John Frere's 1797 speculation that the handaxes he had found came from 'a very remote period indeed; even beyond that of the present world' was elaborated during the following hundred years into a deep and complex prehistory, one with a succession of warm and cold stages, a sequence of stone tool industries, and the beginning of a fossil record of human evolution that conceivably stretched back half a million years or more. Barely two hundred years after Frere's discovery, recent finds from Suffolk have taken the human story in Britain back to an even more remote time, one that is unquestionably beyond that of the present world.

CHAPTER ONE

THE FIRST BRITONS

PRECEDING PAGES: 700,000 years ago the landscape of East Anglia was dominated by rivers flowing eastwards through forests towards the ancestral North Sea. These ancient river courses provided routes across the landscape for the first Britons.

We now believe that the early humans represented by Eugene Dubois's finds from Java – known as *Homo erectus* – had evolved in Africa and started to spread from there nearly two million years ago, the first species definitely known to have done so. Initially these early humans were thought to have spread eastwards to China and Indonesia, keeping to subtropical and tropical environments that were familiar to them. But from 1991 new finds started to be made in western Asia, in Georgia, at a site called Dmanisi. The site lay under a ruined medieval village, but the archaeologists digging it were puzzled to find a rhinoceros tooth in the wall of the cellar in one of the buildings. Palaeontologists identified this tooth as an early Pleistocene fossil, and further excavations showed that a much more ancient site made up the hill on which the village had been built.

This work yielded many more fossils, including scimitar-toothed cats and other large predators, as well as giant ostriches. It also produced a chinless human jawbone. Georgian scientists and American collaborators argued that the site was nearly two million years old, which many experts found difficult to accept. However, further research has backed up these claims, placing the date at about 1.7 million years, and producing five small-brained human skulls, three more jawbones, and many other parts of the skeleton. So primitive humans were living in western Asia, close to the eastern borders of Europe. The stone tools they were using were as basic as those made by their presumed African ancestors from two million years ago – sharp but simple pebble and flake tools, made with one or two blows of a hammerstone.

If people were in western Asia over 1.5 million years ago, why did they leave Africa, and how and when did they first arrive in Europe? It used to be thought that the first move out of Africa must have been fuelled by changes in behaviour, bigger brains, or better tools, but it is difficult to discern such developments in the evidence preserved at Dmanisi. Certainly some of the animals there, including large carnivores and giant ostriches, had also spread from Africa, and it may well be that the Georgian environments were not that different from those of North Africa. Additionally, two of the large African-derived carnivores were scimitar-tooth cats, and these specialized animals lacked the teeth to strip a carcass clean of its meat, or break the thicker bones of their prey. Ancestral humans in Africa are thought to have undergone a change in physique about two million years ago, evolving a more human body shape, with longer legs powered by reshaped muscles and tendons, and a well-defined neck and waist. It has been suggested that this physique evolved to help early humans get to dead animals before the scavenger competition by endurance running. Humans cannot run very fast, but we are (when fit!) steady but supreme long-distance runners, and we can envisage early *H. erectus* scanning the horizon for the first sign of vultures circling a dying or dead animal, and then making off towards it. Getting ahead of the hyaenas and jackals and fending off the vultures would have guaranteed a good supply of meat, whether the carcass was a natural death or the victim of a predator.

Medium-sized social carnivores such as hyaenas and wolves have large hunting and scavenging ranges, and as archaeological evidence suggests that

ABOVE: The most ancient human site in Eurasia lies on a hill underneath a later archaeological site – the medieval village of Dmanisi.

human ancestors in Africa increasingly switched to carnivory about two million years ago, perhaps this alone was enough to push the expansion of early human ranges out of Africa. Some experts have calculated that a switch to meat and an increase in body size documented in the fossil record at this time would have necessitated a ten-fold increase in feeding range compared with their supposed African ancestors. While dependence on particular plant resources would have limited expansion, switching to meat eating meant that humans could follow the herds and even switch food species as the need arose.

Three ancient routes into Europe have been proposed: the most obvious one is via western Asia – either from the Caucasus (including modern countries such as Armenia and Georgia) or from the Levant, the corridor connecting north-east Africa and western Asia (including countries such as Israel and Syria). Two other possible and more direct pathways from North Africa into southern Europe have been proposed, via routes across the Mediterranean. The more westerly one might have been from what is now Morocco into what is now Spain or Gibraltar, while a central one might have led from present-day Tunisia to Italy via Sicily. These last two routes would have been easier if the sea level was affected by the periodic growth of the ice caps, lowering the Mediterranean and exposing more land, with additional islands between. However, there were never continuous land bridges between Africa and Europe in these two regions, so any early human pioneers contemplating such a journey would have required a raft. So most experts believe that the eastern route into Europe from western Asia was by far the most viable and probable, and this region does have evidence of human occupation from over a million years ago such as Ubeidiya in Israel and Dmanisi, already mentioned. In addition, there is evidence of the spread of mammals that had originated in Africa from western Asia into Europe such as large predators and a baboon-like monkey. But comparably early archaeological evidence from Europe is either absent or at best highly contentious. It has been claimed that artefacts from sites in France and southern Spain date from over a million years ago, but many experts are doubtful of the artefacts, the dating, or both. So when did humans first arrive in Europe, and in Britain?

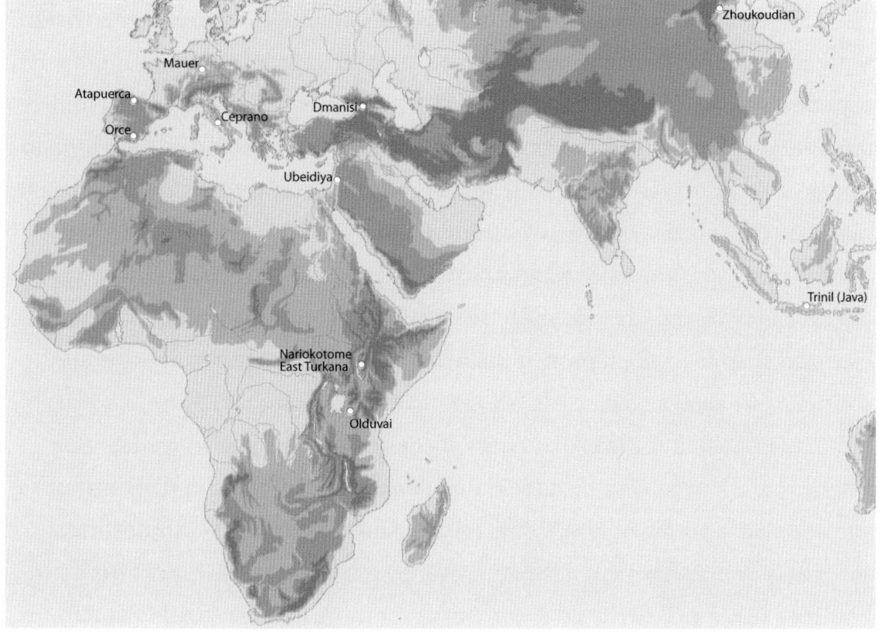

LEFT: Some of the earliest sites with human fossils or stone tools.

Many archaeologists also believed that people then were poor hunters, and would have only been able to scavenge, rather than hunt, anything larger or fiercer than a rabbit

The general belief has been that humans, emerging from their African homeland, would have taken a long time to find ways of coping with the very different environments of Europe, and in particular northern Europe, where average temperatures were lower, growing seasons and winter days shorter, and winter conditions much more severe. Behavioural adaptations such as skin clothing, making shelters and building fires, would all have helped in the north, but archaeological evidence suggests that these innovations came later in the European record. Many archaeologists also believed that people then were poor hunters, and would have only been able to scavenge, rather than hunt, anything larger or fiercer than a rabbit. Thus early humans would have had to loiter in the background while lions, hyaenas, vultures or wolves got their fill, and then were left to scrape the bones and break open the most inaccessible parts of the carcass with their stone tools to extract any remnants of marrow. With fierce competition from a range of formidable carnivores, humans would have needed much better technology and social organization before they were able to survive in areas like Britain, or so it was argued.

The 1935 discovery of the Swanscombe skull in a Kent gravel pit had been a landmark for archaeology and human fossils in Britain, particularly as Piltdown edged ever closer to the palaeontological dustbin, but it also seemed to confirm humans in Britain as definite latecomers. Britain was at the end of the migration routes of those first colonizers: the Mauer jaw suggested that people might have reached as far north as Germany during a first wave of advance about half a million years ago, but there was no evidence that they had reached Britain until after the massive ice advance called the Anglian, dated about 450,000 years ago (see next chapter), which pushed the River Thames southwards to its present course. As we shall also see, things began to change in 1993 with the discovery of human remains at Boxgrove, near Chichester, that seemed to predate the Anglian ice advance. This suggested that people actually were in Britain about 500,000 years ago, around the same age as Mauer. The records show us what animals they may have eaten, what tools they used and, in the case of Boxgrove, how they made and deployed them. But as for more details of their adaptations, for example whether they had camps, or used fire, the sites are silent – the evidence is simply not there.

Were these, in fact, the oldest records of humans in Europe? That certainly was the predominant opinion about ten years ago, encapsulated in the views of two Dutch scientists, Wil Roebroeks and Thijs van Kolfschoten, in what became known as the Short Chronology. In summary, the argument went that from 500,000 years ago there was clear evidence of a human presence in the form of well-dated artefacts such as the handaxes from Boxgrove, well-stratified butchered bones and, most unequivocally, the human fossils from Boxgrove and Mauer. But before that, the evidence was either sparse or poorly dated, and in the case of human fossils, non-existent. Diligent searches for over a century in the pre-Anglian deposits of Norfolk and Suffolk had yielded up many thousands of fossil mammal bones, a smattering of dubious eoliths, but no credible butchered bones or human fossils. Good evidence only appeared just before the Anglian glaciation of about 450,000 years ago, not in East Anglia itself, but further south at Boxgrove and, perhaps, the lowest levels of Kent's Cavern, dug so long ago by MacEnery and Pengelly, dated by their association with fossils of a water vole called *Arvicola terrestris cantiana*, as was also the case at Mauer. Roebroeks and van Kolfschoten argued that this was the situation throughout Europe. Humans may have evolved in Africa by two million years ago, and spread eastwards to China and Indonesia by a million years ago, but European conditions were much more challenging and less suited to a primate of tropical origins, at least until human adaptations were finally up to speed in coping with those challenges, something not achieved until about half a million years ago.

But the Short Chronology has since had to give way twice, first in southern Europe, and then, with the work of AHOB, further north as well. There is now strong evidence that humans were in fact in southern Europe much earlier than 500,000 years ago. In southern Spain, at Orce, some archaeologists believe they can date stone tools as far back as 1.5 million years, but these claims remain highly controversial. From a later date, a fossil human braincase with large brow ridges, resembling both *H. erectus* and *H. heidelbergensis*, has been discovered in deposits on a hillside in central Italy at Ceprano. It derives from levels which elsewhere contain simple stone tools (no handaxes), with an approximate date of 800,000 years. From a similar age, but with somewhat better dating, the

Atapuerca Hills in northern Spain have produced fragmentary human fossils of several adults and children. This is an open site called Gran Dolina, exposed by an old railway cutting through the Sierra de Atapuerca, and it is filled with earthy sediments containing bones and stone tools. The layer containing the human fossils and associated stone tools (again simple ones, not handaxes) lies immediately under a level that records the last time that the Earth's magnetic poles underwent a major switch in orientation (this is discussed further in the next chapter) – a switch which has been dated at about 780,000 years ago. Furthermore, these human fossils were definitely associated with the primitive vole *Mimomys*, rather than its descendant *Arvicola*, so something known as the vole clock signalled a greater age than Boxgrove or Mauer. The vole clock is based on the evolutionary transition between the primitive and extinct vole called *Mimomys savini* and the water vole *Arvicola terrestris cantiana. Mimomys* had molar teeth with closed roots, whereas the molars of its descendant, *Arvicola*, are open-rooted and ever growing. Early sites in Europe (such as Orce and Gran Dolina) have *Mimomys*, while later sites (like Mauer and Boxgrove) have *Arvicola* fossils.

OPPOSITE: This partial child's skull from Gran Dolina (Atapuerca) in Spain dates from around 800,000 years ago.

The Gran Dolina fossils include the bones of the forehead and face of a child, fragments of jaws, teeth, arm and foot bones, and even a kneecap, and most show marks that suggest they were cut by stone tools. While we might have attributed such damage to burial practices if we were dealing with recent humans and their complex behaviour, this seems unlikely for early humans, leading to the suggestion that these individuals were the victims of ancient cannibals, whether of their own or another group. Early studies of the fossils led the Spanish workers who have studied the bones to propose that they represented a new species of human called *H. antecessor*, Pioneer Man. They originally argued that this species evolved in Africa from *H. ergaster*, where it went on to give rise to *H. sapiens*, while the population that spread to Europe eventually gave rise to the Neanderthals. Thus *antecessor* was the last common ancestor both of our species and the Neanderthals. However, some of the team have recently rethought these ideas and now believe that features in the face, jawbones and teeth may indicate that *antecessor* is more closely related to fossils of *H. erectus* from China. They argue that an alternative scenario has *antecessor* in Europe, with its basic tools, being replaced by populations of *heidelbergensis*

ABOVE: This picture taken in March 2006 shows how erosion at Happisburgh in Norfolk is destroying houses and roads at an accelerating rate.

that originated in Africa and spread from there, bearing handaxe tools. It is these people who would then have been the ancestors of the Neanderthals, not *antecessor*, which died out.

Finds from Italy and Spain thus successfully challenged the Short Chronology, but one of its originators, Wil Roebroeks, argued that the model still held true for northern Europe, where good evidence of human occupation before 500,000 years ago, and before *Arvicola*, was still lacking, despite nearly 200 years of careful searching in regions such as East Anglia. But it is there that new and critical evidence has finally emerged. This breakthrough has come about because of an unrelenting and, for some people, catastrophic advance of the North Sea across parts of the coasts of Norfolk and Suffolk. The last fifteen years have seen a dramatic increase in this erosion, with arguments raging over the causes and what can be done about it. Greatly increased dredging for sand and gravel offshore, improved coastal defences in some regions deflecting erosion to other areas, global warming increasing the frequency and the strength of storms – all these have been implicated at times, and all may indeed be contributing. But whatever the reason, an old Norfolk coastal village called Happisburgh (local pronunciation Haysborough) has seen its cliffs cut back viciously with houses, roads and the lifeboat ramp lost to the sea, and the creation of a new bay over 500 metres wide. This tragedy for the local inhabitants has been accompanied by the exposure of long-buried deposits in and under the cliffs that date from before the Anglian ice advance of about 450,000 years ago.

A few years ago, in a muddy deposit on the foreshore exposed only at low tide, a local archaeologist uncovered a beautiful handaxe made of black flint, and bones of deer and bison. By then, AHOB researcher Simon Parfitt had studied a collection of mammal bones purchased by the Natural History Museum in 1897, including some from Happisburgh. The collection was made by one of the greatest accumulators of fossil evidence from East Anglia – the geologist Alfred Savin, after whom several of the species represented have been named, including the vole *Mimomys savini*. One find was particularly significant as it was the foot bone of a bison, and it had multiple cut marks on it, marks that had not been noticed over the previous hundred years when it had sat in a museum drawer.

ABOVE: Local resident Mike Chambers found this beautiful black flint handaxe at Happisburgh while beach walking in March 2000.

With the help of many dedicated local collectors, AHOB has gathered much more evidence from Happisburgh and, despite the cold and the tides, has carried out three excavations there, showing that this was indeed a site of ancient human occupation at least 500,000 years ago. Several other large bones show cut marks and impact damage, suggesting they were processed for meat and marrow, and this is backed up by the recovery of more than fifty flint flakes and a second handaxe. Over seven tons of excavated sediments have been collected and washed through fine sieves either on site or back in the museum, producing rare remains of smaller animals, including fish, birds and amphibians, with the only British record of an extinct kind of frog called *Pliobatrachus*. The clayey sediments also contained well-preserved remains of plants such as pollen, seeds and wood, and of beetles, suggesting that the artefacts and bones were deposited in the backwater of a large river bordered by coniferous woodland, near the peak of an early Middle Pleistocene warm stage (interglacial). The river system will be discussed later, but which interglacial was it? Here there are two somewhat contradictory interpretations of the evidence

within the AHOB project. Amongst the small mammal remains found in the sieves are teeth of *Arvicola*, suggesting that the site is about the same age as Boxgrove, assuming the overlying deposits in the cliffs are debris from the Anglian ice sheet advance. But there is another possibility. Jim Rose of AHOB believes that the Anglian was not, as commonly believed, the first ice sheet to push down across East Anglia. He and colleagues argue that there was an earlier ice advance about 650,000 years ago, which came down the British mainland rather than across from Scandinavia, picking up a distinctive Scottish suite of rocks as it did so. Jim argues that the deposits above the Happisburgh handaxes and bones show the distinctive geological signatures of the earlier ice advance – in fact Happisburgh has been designated as the type-site of this newly recognized glaciation. In which case the evidence of human occupation at Happisburgh would be an astonishing 700,000 rather than 500,000 years old. However, other workers, within and outside AHOB, dispute this view.

This ongoing debate about the geology of East Anglia is critical to the dating of many sites beyond Happisburgh, and before we move on to discuss another and even more important site it is worth spending a little time looking at what we know about the quite different landscapes and geography of the region more than half a million years ago. Britain was then a peninsula joined to the Eurasian continent by a wide land bridge, whose spine was a chalk ridge running between what is now the south-east of England, and France. Many of today's river systems, for example the Severn and Bristol Avon, did not exist. Others were on different courses, such as the Thames, which flowed along a more north-easterly course through the Vale of St Albans towards Clacton and Colchester. Rivers also flowed from the Pennines and the Midlands that have since completely disappeared – for example, the Mathon River flowed southwards through Herefordshire, and a mighty river called the Bytham, 300 kilometres (200 miles) long, flowed south and then eastwards across East Anglia. Several important archaeological sites lie along the route of the lost Bytham, including Waverley Wood near Coventry, which has produced early handaxe tools, and AHOB sites High Lodge and Warren Hill near the border between Norfolk and Suffolk. The Waverley Wood handaxes are distinctive because they are made of quartzite and the volcanic rock andesite, rather than flint, but the workmanship is every bit as skilful as in more typical examples. When the ice

of the Anglian glaciation advanced to its maximum extent, it obliterated many of the old rivers, and pushed the Thames southwards. At the height of the cold stage a massive front of ice sat over the Midlands and north London, and south of it were huge lakes filled by spring thaws and ice-dammed rivers, such as the Thames. One known as Lake Harrison spanned Leicester, Coventry, Rugby and Leamington, and when another eventually burst its banks near Hillingdon, just west of London, a catastrophic flood freed the Thames, where it took over river valleys such as the proto-Medway to form at last its present estuary.

ABOVE: These rocks were placed under the cliffs at Happisburgh in 2003 to protect them from the advancing sea, but they now mark the alarming extent of erosion since then.

Dating before these ice age dramas, High Lodge, near Mildenhall, remains an enigmatic site. Despite 150 years of excavations, its extraordinary history is only now becoming clear. Workmen who were digging clay for the manufacture of bricks discovered the first artefacts in the 1860s, and ten years later John Evans (one of de Perthes' supporters, who we encountered in the previous chapter) noted that tools looking Middle Palaeolithic seemed to underlie ones that were Lower Palaeolithic, an inversion of the expected succession. Several excavations followed over the next ninety years, and all found the same peculiar sequence of

ABOVE: Over a century after it was excavated at Happisburgh, Simon Parfitt recognized cut marks produced by a stone tool on this bison foot bone – the first indication that ancient humans had been at the site.

artefacts. The most recent pre-AHOB excavations found evidence from fossils and overlying glacial deposits to show that the site was pre-Anglian, and our subsequent work has recovered more stone tools, fossil mammals such as the extinct giant beaver, and large chunks of wood. Both the artefact assemblages are indeed pre-Anglian, so the 'Middle Palaeolithic' ones, despite their refinement, are clearly very ancient. And the deposits that contain these artefacts were probably originally from an interglacial lake or river that was frozen solid in the subsequent Anglian ice age and was bulldozed miles across Norfolk enclosed within glacial debris as a gigantic glacial erratic. The nearby site of Warren Hill has long been known as a site for handaxes, perhaps the most prolific one in the country. Our excavations exposed a shallow marine or tidally deposited sand at the base of the sequence, without artefacts, and upper deposits from a southward-flowing part of the Bytham River system containing handaxes, and knapping debris from their manufacture. We also found large heaps of a fine gravel spoil, apparently the dumps of collectors sieving for handaxes more than a hundred years ago.

As the Bytham River slowed past Warren Hill towards its delta on what is now the East Anglian coast, it deposited sediments on the edge of a huge north-facing bay, into which the Rhine also flowed. The sites of Norton Subcourse in Norfolk and nearby Pakefield, just over the border in Suffolk, were probably both related to the Bytham, and they record a time when the climate of Britain was balmy and Mediterranean, and this part of East Anglia was a fertile estuarine plain. Norton Subcourse has a working quarry for sand and gravel, and in its depths are rich clays, silts and peats, containing evidence of a fen surrounded by reeds and alder. AHOB excavations have found the bones of amphibians, reptiles and fish such as carp and pike, and of mammals such as an extinct ass-like horse, elephant, deer and hippos, with remains of the latter showing gnaw marks on their backbones. Nearby, piles of bone-rich hyaena droppings lie just as they were deposited some 700,000 years ago, and we can imagine the hyaenas climbing on to the backs of dead hippos in the swamp in order to feed. Amongst the small mammal remains are teeth of *Mimomys*, confirming that Norton is indeed older than Boxgrove and Happisburgh.

About 30 kilometres (20 miles) downriver, on the Suffolk coast near Lowestoft, is Pakefield, and this small town has now entered the archaeological

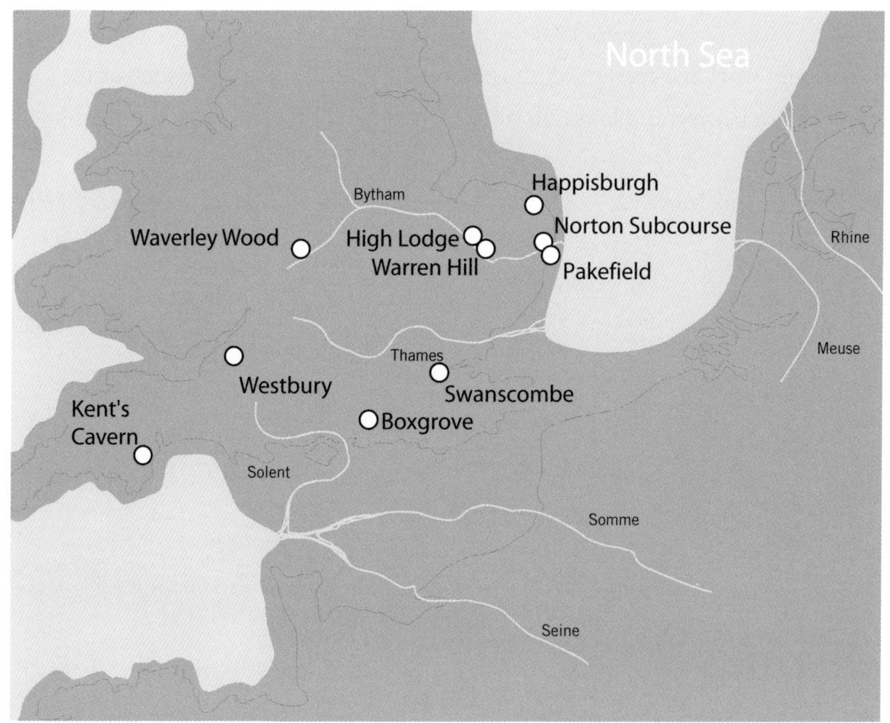

LEFT: A reconstruction of the geography of Britain during the earliest periods of human occupation.

OVERLEAF: Hippos amongst dense delta vegetation in Botswana. East Anglia must have witnessed similar scenes 700,000 years ago.

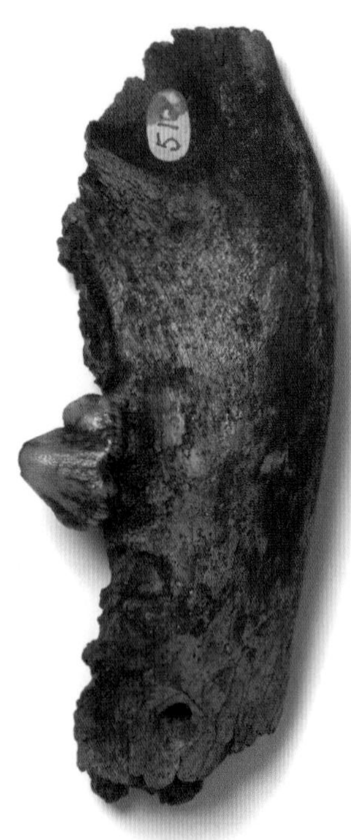

ABOVE: A 700,000-year-old hyaena jaw from Pakefield in Suffolk.

OPPOSITE: Some of the ancient but still razor-sharp flint flakes found at Pakefield.

hall of fame, sitting alongside Boxgrove, Swanscombe and, at a much later date, Stonehenge. The site, like Happisburgh and so many others in East Anglia, lies under a cliff of glacial deposits, and for over a hundred years has been producing fossil bones of animals such as hippo, rhino, primitive mammoth, straight-tusked elephant, bison, three species of extinct giant deer, and carnivores such as scimitar-toothed cat, lion, hyaena, wolf and bear. Also there was the primitive vole *Mimomys*, rather than its descendant *Arvicola*, suggesting that this was a relatively old site in the East Anglian sequences. But in the last few years, field visits and local collectors have reported finding flints that looked as if they were struck off by humans, and that has made the site of particular interest to AHOB. Expert examination confirmed that these were indeed humanly worked, but did they really come from the muddy layers under the glacial deposits that were producing the fossil mammals, or could they have been washing down from the fields above the glacial deposits? There was only one way to settle it, and so AHOB worked with the local collectors to carry out further excavations and publish the evidence. These put any doubts to rest and there are now thirty-two worked flints including a core and several retouched flakes. The artefacts are in sharp condition and are made of good quality black flint that was probably picked up in the form of river cobbles. The oldest was found in river estuary deposits where, along with the discovery of rare bones of dolphin and walrus, associated microfossils suggest brackish water. The remaining artefacts were found higher up in dark sediments from the riverbed or riverbank.

Pakefield has produced other valuable material in the form of incredibly rich remains of wood, plants, beetles, molluscs and microfossils, all of which can help paint a vivid picture of the area when the first Britons were there, including very detailed climatic information. There are over 150 plant species, including water chestnut, floating fern, brittle waternymph and broom crowberry, suggesting warmer summers than today. Thousands of beetle fossils represent nearly a hundred different species, including exotic diving species and others that live in rotting wood. Mutual Climatic Range analysis of the beetle assemblage looks at the tolerances of the different species today in order to work out what kind of climate would have been compatible for all of them, and this also suggests July temperatures between 18°C and 23°C, compared with 15°C

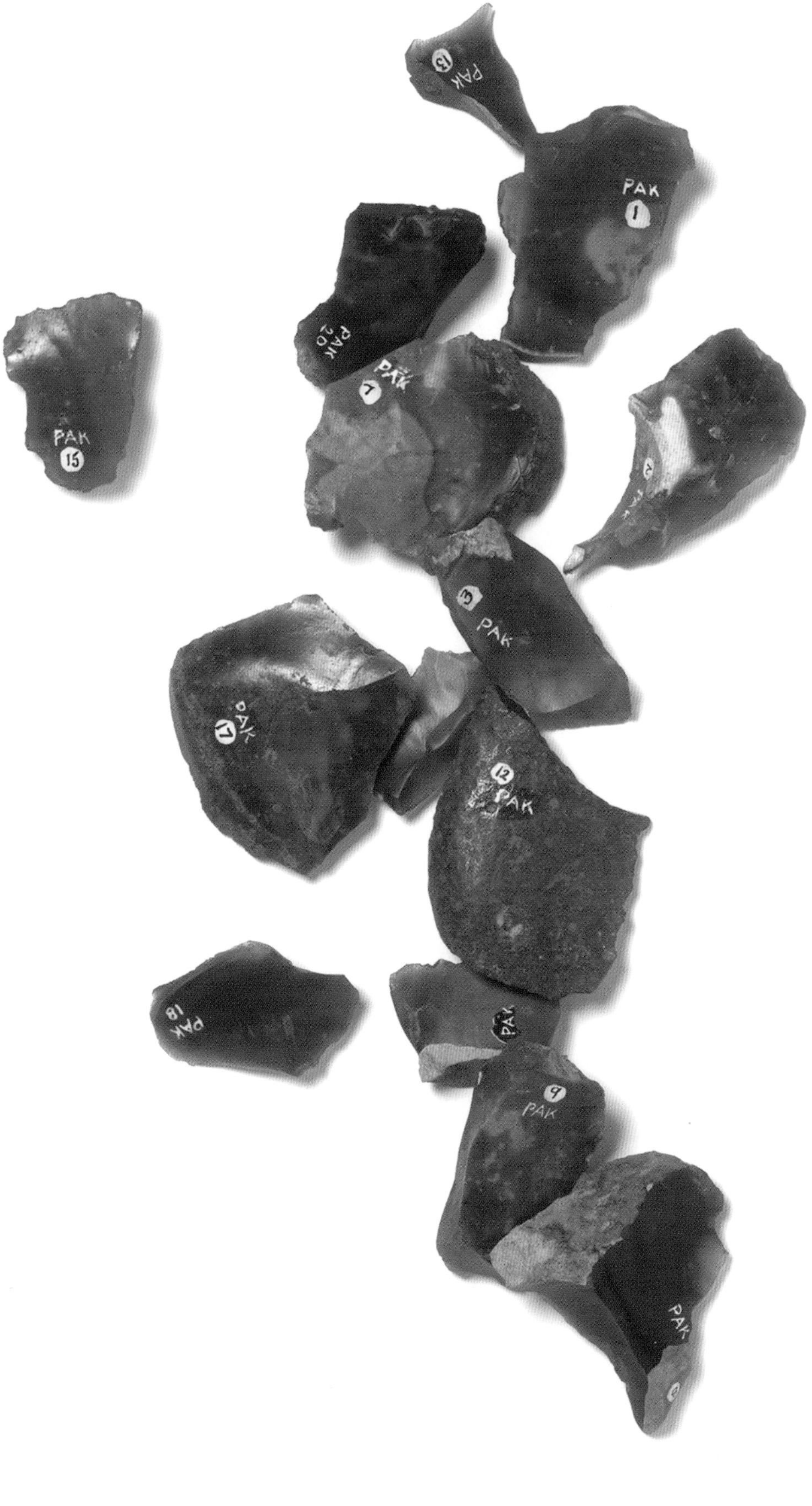
PAK
1
PAK
15
PAK
20
PAK
3
PAK
17
PAK
12
PAK
18
PAK
9

ABOVE: Beetle remains from Pakefield provide important clues to the environment and climate when early humans lived there.

today, while the coldest months (January/February) were mild. In fact we know from the plants, the presence of animals like hippopotamus, and the species of fish and amphibian that winter temperatures were above freezing. And we can even reconstruct rainfall patterns, as the sediments contain carbonate nodules which developed in the soil and which preserve a chemical snapshot of the local climate at the time they formed. Study of the proportions of oxygen and carbon isotopes in these nodules give AHOB researchers a window into ancient climates. The soil carbonates indicate an annual moisture deficit, whilst their isotopic composition reflects intense soil moisture evaporation during their formation, so there was strong seasonal rainfall. The combination of warmer and drier summers and cool wet winters indicates a Mediterranean climate, something unique for a British archaeological site.

The insects, plants, molluscs and other fossils can also tell us about the landscape, suggesting marshy ground with reeds and alder trees near a meandering and extensive river estuary, with pools of different depths, while oak woodland and open grassland grew farther from the river. This mixture of landscapes would have supported the many different browsing and grazing

> The combination of warmer and drier summers and cool wet winters indicates a Mediterranean climate, something unique for a British archaeological site

mammals known from Pakefield, from the size of elephants and hippos down to deer, and also would have attracted their predators and scavengers, from lions to foxes. The floodplain provided an environment rich in plant and animal resources for the humans as well, and most importantly in an area without much rock there were flint-rich river gravels for stone toolmaking. This knowledge of the setting makes interpreting the stone tools difficult, though. The simple form of the core and flakes might indicate a primitive pre-handaxe technological tradition, in line with interpretation of the tools at Ceprano and Gran Dolina. However, the Pakefield tools are few in number, and were being made from water-worn pebbles, probably not suited for the manufacture of large flaked tools like handaxes. But are there other clues to the age of the artefacts at Pakefield?

As at Happisburgh, the river sediments containing the tools at Pakefield are stratified below glacial debris and outwash deposits, so they must date from before the Anglian cold stage of about 450,000 years ago, at least. Jim Rose and colleagues argue that the deposits sandwiched above the tool-bearing layers and below those of the Anglian (high in the cliff) show at least two cycles of high sea levels, with cold periods between, which must indicate at least two separate periods of small ice caps (and therefore interglacials) *after* the Pakefield interglacial and *before* the cold of the Anglian. They would place the Pakefield interglacial at least as far back as 700,000 years, and perhaps even earlier. Others in AHOB are not sure about this geological approach, and instead rely on the fact that the interglacial at Pakefield is warmer than any of the 'Cromerian' (after Cromer in Norfolk) interglacials so far known and dated at 500,000–600,000 years, and on the presence of two important species of the vole *Mimomys*. One is the familiar species *Mimomys savini*, placing the site as older than Boxgrove, Mauer and Happisburgh, all of which have the descendant form *Arvicola*. A second *Mimomys* species, *M. pusillus*, is extremely rare in Britain and is not

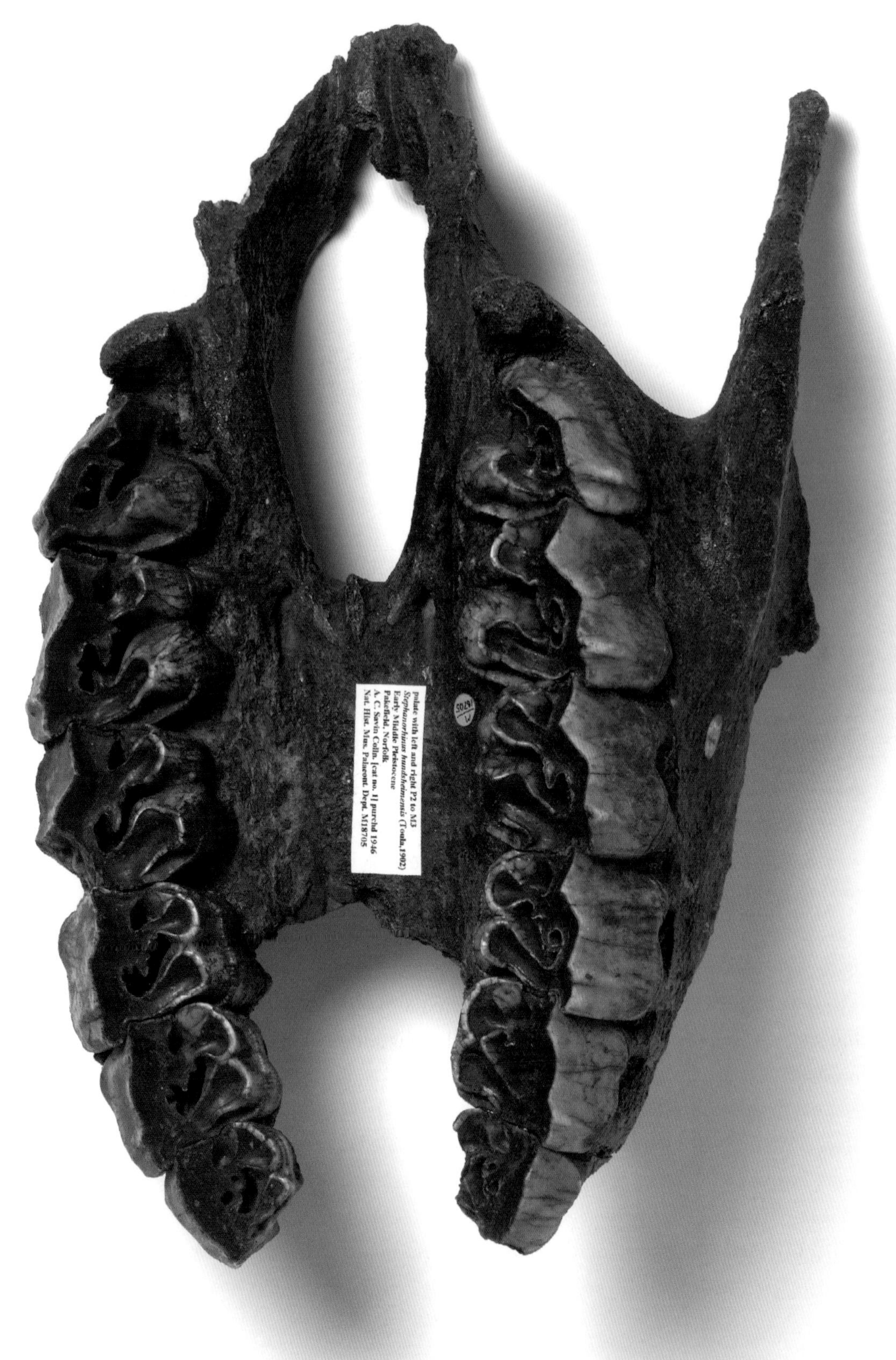
palate with left and right P2 to M3
Stephanorhinus hundsheimensis (Toula,1902)
Early Middle Pleistocene
Pakefield, Norfolk
A. C. Savin Colln. [cat no. 1] purchd 1946
Nat. Hist. Mus. Palaeont. Dept. M18705

known anywhere in the world after about 650,000 years ago. For our purposes, the important fact is that the geological, climatic and vole clock methods all date the stone tools from Pakefield as far back as 700,000 years ago. And there is one more bit of science that backs this up.

Amino-acid dating is based on the principle that the twenty amino acids that make up living proteins start to change once an organism dies. Most amino acids in proteins have one or more asymmetric carbon atoms and in living things these extend to the left side. However, for each one, there is an alternative isomer where the asymmetric carbon atoms extend to the right side instead (these are like mirror images of each other, just as right and left hands are). The isomers can be distinguished by whether they bend a beam of polarized light to the right or left, and the interesting thing is that after its death, in a process called racemization, the proteins of an organism start switching to their isomers at a steady rate dependent on temperature and time. Thus the more racemization that has occurred (all other things being equal), the longer the time since the organism has died. Amino-acid dating has been used on many different fossil materials including human and animal bones and teeth, plants, eggshells and shellfish. The technique has undoubtedly had a chequered history as a dating method, but AHOB associate Kirsty Penkman has used a new and refined procedure on shellfish that lived in the ancient river muds at Pakefield, and she compared results with other British sites. Those from Pakefield show a high degree of racemization compared with sites with *Arvicola* commonly dated to 400,000 and 500,000 years ago, and are similar to or slightly more racemized than Cromerian sites with *Mimomys* dated at about 600,000 years. And we know that none of these interglacial sites were as warm as the Pakefield interglacial, so that one has to be older.

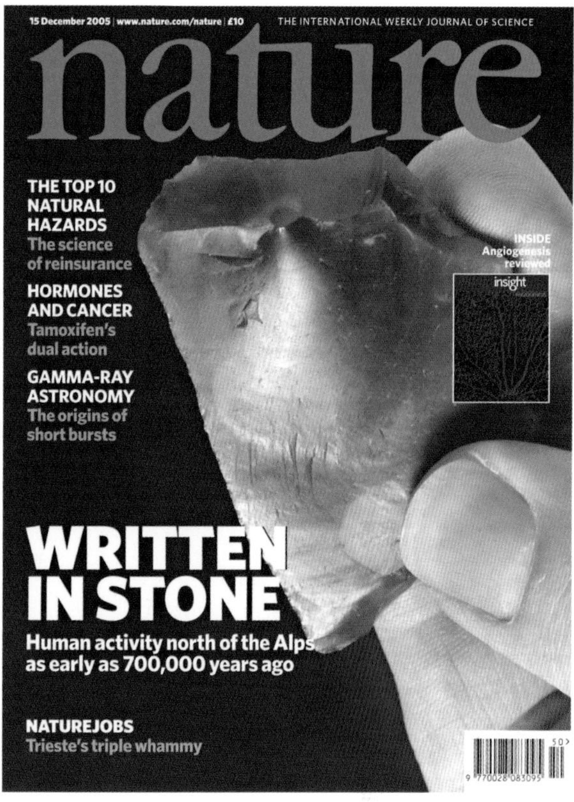
15 December 2005 | www.nature.com/nature | £10
THE INTERNATIONAL WEEKLY JOURNAL OF SCIENCE
nature
THE TOP 10 NATURAL HAZARDS
The science of reinsurance
HORMONES AND CANCER
Tamoxifen's dual action
GAMMA-RAY ASTRONOMY
The origins of short bursts
WRITTEN IN STONE
Human activity north of the Alps as early as 700,000 years ago
NATUREJOBS
Trieste's triple whammy

ABOVE: The oldest evidence of human presence in northern Europe (Pakefield) was published in December 2005.

OPPOSITE: The beautifully preserved upper jaw of a rhinoceros from Pakefield.

Geology, climate, voles and amino-acids together place the stone tools from Pakefield at least 700,000 years ago, making them by far the oldest evidence we have for people in Europe north of the Alps. As we have seen, there has been much discussion about what additional social, technological or bodily adaptations humans would have needed to colonize north-west Europe compared with their occupation further south, but the Mediterranean climate reconstructed for the archaeological levels at Pakefield implies that these pioneers spread northwards in familiar climatic conditions. In accepting the stone tool evidence from Pakefield that required him to modify his Short Chronology model still further, Wil Roebroeks pointed out that the palaeoenvironmental evidence showed that the tools were discarded along the shores of an early Middle Pleistocene 'Costa del Cromer'. In this sense the Pakefield evidence still supports the Short Chronology in that this was probably a brief episode of rapid migration north under favourable conditions, really requiring little more of people than what was needed in Italy or Spain. Substantial occupation of, and adaptation to, northern Europe probably still only happened at the time of Mauer and Boxgrove. But if even within England, Roebroeks added, one of the best and longest researched parts of the world, surprises like Pakefield could turn up, what else was still to come? Even as this book is being written, there certainly promise to be yet more revelations from the sites of East Anglia . . .

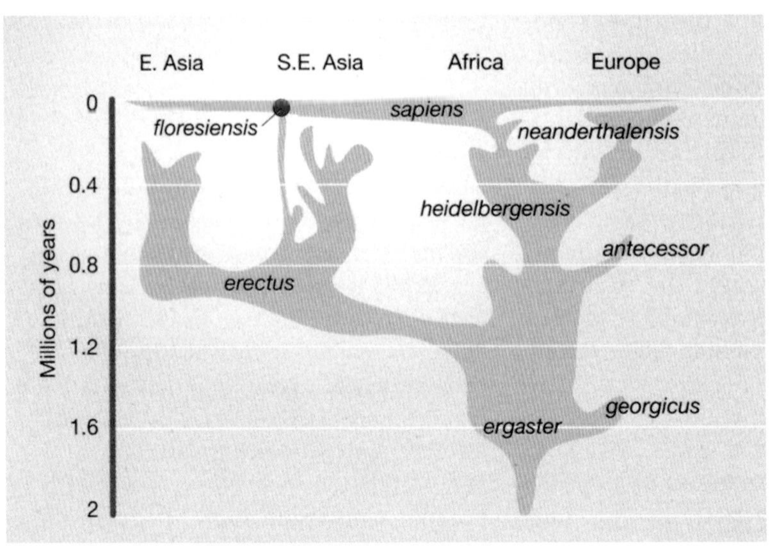

RIGHT: A representation of the later stages of human evolution by Marta Lahr and Robert Foley. The Dmanisi fossils are sometimes referred to the species '*Homo georgicus*', while *Homo floresiensis* is the controversial species colloquially known as the 'Hobbit'.

AHOB has opened an entirely new window on the first Britons, pushing their date back 200,000 years beyond Boxgrove and 300,000 before Swanscombe, and in doing so has emphasized how long our prehistory is and how little we still know about it. It has also completely changed our view of the early colonization of Europe, as we now know that humans had reached the north-west edge of the inhabited world (apart from Ireland, which always remained out of reach in the Palaeolithic) some 700,000 years ago. The routes they took into Britain are unknown, but there were probably at least two. If they followed the valley of the ancient Rhine westwards they would have arrived at the huge north-facing bay into which the proto-Thames and Bytham also flowed, and their valleys would have provided good routes into the British peninsula. If they came from the south they could have followed the valley of the Somme until it reached the Atlantic, and then crossed the chalk ridge into southern Britain. From there, they could have travelled west along the coastal plain or northwards up river valleys such as the proto-Solent, or the precursors of the Vales of Gloucester and Evesham. Up to now, we have not mentioned that most significant part of our present geography, the English Channel, because it did not exist then, but its development will loom large later in the book. Before we get to that part of the story, we will look in the next chapter at the human colonization of Britain, when humans of the species *Homo heidelbergensis* arrived. We will also examine how the evidence to date these early arrivals has been pieced together from records on land and from deep in the oceans.

CHAPTER TWO

UNDERSTANDING ICE AGES

PRECEDING PAGES: Sunlight on a glacier in the Antarctic. Many miles of coastline would have looked like this in Britain 450,000 years ago.

In the last chapter we saw how new discoveries have changed our views of the first peopling of Europe, and Britain. But although we can build a good picture of the landscape and the climate, the evidence of human behaviour itself is fragmentary: from the glimpses we have at places like Atapuerca and Pakefield, it is difficult to build much of a picture of what life was like for these European pioneers. Moving on in time, however, the evidence is suddenly much richer: we can point to a spot in the ground a few miles from the city of Chichester in Sussex and confidently say that this was where someone squatted down to make a flint handaxe about half a million years ago, and even show how they made that handaxe. We can also reconstruct in detail how these tools were used to expertly fillet meat from the carcass of a horse, a deer or a rhino. I'll now unravel how we know these things, and how we know the age of the ancient sites in which this evidence is preserved.

I explained in the Introduction that it was gradually realized from geological evidence in the Alps that there had been a succession of ice ages, and how it was not just in Britain that seeking the local equivalents of these Alpine ice ages became something of an obsession. They were regarded by many workers as geological universals that were waiting to be identified in every continent of the world, and this often impeded progress rather than helped it, by providing straitjackets of conformity even where things looked very different. What was needed to break out of the straitjacket were much more detailed and continuous archives of past climatic changes than could be found in the fragmentary sequences of the Alps or the Scottish highlands, that could help translate past climatic records across the whole planet. And just as important as finding better records was the need to know what lay behind the ice ages. As with the model of continental drift, which languished in neglect until geological data were discovered which showed that the continents really had moved around, an explanation for the ice ages was proposed in 1924, but it was fifty years before its veracity began to be generally recognized.

A Serbian mathematician called Milutin Milankovitch used very precise calculations of the orbits of the planets to model the position of the Earth in relation to the sun over the last 600,000 years. The Earth's orbit is not exactly circular, so the time of the year when it is furthest from the sun (and thus receives less heat) changes through time, over many thousands of years. In addition, two different factors affect the tilt of the Earth's axis of rotation in relation to the sun through time, accentuating or reducing the differences in the seasons. From these data Milankovitch was able to reconstruct the varying patterns of summer warmth in the northern hemisphere and show how the polar ice caps would have waxed and waned in the face of hotter or cooler summers. If all three factors (orbit shape and the two tilt factors) increased the amount of sunlight falling on the Earth, there was an interglacial. If the factors all worked in the opposite direction and sunlight was reduced, the Earth suffered an ice age. Consequently, he was able to identify three periods of particularly low solar warmth that might correspond to the Gunz, Mindel and Riss glacials, as well as a prolonged period of warmth that could have corresponded to the Mindel-Riss or Great Interglacial of Penck and Bruckner. But Milankovitch realized that most of the time the three factors, working on

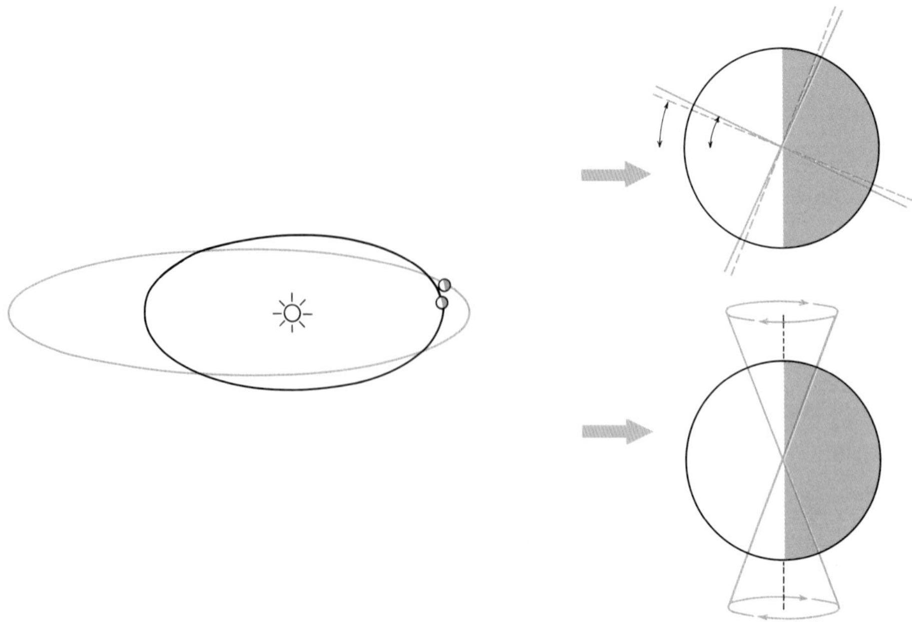

ABOVE: The Milankovitch cycles are a major factor in producing climate change. They relate to (*left*) the eccentricity of the Earth's orbit (100,000-year cycle); (*top right*) the differing tilt of the Earth's axis (41,000-year cycle); and (*bottom right*) precession as the Earth wobbles slightly as it spins (23,000-year cycle).

cycles of about 100,000, 40,000 and 20,000 years, would be pulling in different directions, so the reality of climate change was going to be a lot more complex than what had been observed in the Alps.

In the 1950s a 'Rosetta Stone' of past climate at last began to be uncovered, not from desert sands but from the ocean floor. The man who began the laborious process of reading the record was an Italian palaeontologist called Cesare Emiliani. Despite the disruption of World War II, he managed to complete his doctoral research on marine microorganisms in Italy, and then moved to Chicago to continue his studies. At that time, the first deep cores were being drilled through the 'calcareous oozes' (accumulations of the chalky shells of dead microorganisms) on the tropical ocean floors, with the aim of looking at the environment of the oceans through time – the deeper the ocean floor sediment, the older the record it contained. Although Emiliani was interested in the species of fossil microorganisms preserved in the deep sea cores, as these might give clues about past changes in the oceans, he concentrated on the chemistry of their carbonate shells or tests. In theory, the composition of the tests the creatures built should reflect the kind of ocean in which they had lived, whether warm or cold, fresher or saltier. In particular Emiliani analysed the ratio of two stable isotopes of oxygen – ^{16}oxygen (the normal kind) and ^{18}oxygen (a rarer and heavier form) – in the tests sampled at 10cm intervals of depth.

Using the theory that more ^{16}O was an indication of warm oceans, and more ^{18}O colder oceans, he was astonished to find that the oxygen isotope ratios in the cores fluctuated dramatically in a saw-tooth pattern, and that even the subtropical Pacific and Caribbean oceans had regularly been several degrees colder in the past. What was more, he identified seven cold stages in the Caribbean cores, and fifteen in the Pacific cores, compared with Penck and Bruckner's four in the Alps. Counting from the top of the cores, Emiliani identified each warm stage by an odd number (our present interglacial is thus Marine Isotope Stage 1, or MIS 1), and each cold stage by an even number (the peak of the last ice age is thus MIS 2).

The next development in unravelling the marine record came from two geologists, Nick Shackleton at Cambridge and the American John Imbrie. They suggested that factors other than local water temperature lay behind the oxygen isotope fluctuations, and these were global. In their view, the main reason behind past oxygen isotope variation in the oceans was the size of the ice caps. Water that evaporates from the oceans is isotopically 'light' – that is, it has proportionately more ^{16}O. Ice caps are made up of atmospheric water that has fallen as rain, hail or snow, or has been directly frozen from the air by low temperatures, like frost. Thus if ice caps grow, they store up more and more light water, and the ocean waters left behind get 'heavier' (more ^{18}O). If ice caps melt, the ^{18}O-rich oceans get diluted with more ^{16}O again. So a low proportion of ^{18}O atoms in the marine core fossils indicates a time of small ice caps and high sea levels (an interglacial), while a high proportion of ^{18}O must indicate large ice caps and a relatively low sea level (a glacial).

To be fair to Emiliani, he was aware of ice cap influence as well, but he considered it much less significant than the local water temperature in its effects on the chemical composition of the microfossils. But with this breakthrough in interpretation, it was at last possible to get a global view of past climatic changes and test Milankovitch's theories. To do this, there had to be a time scale for the isotope changes in the cores, and hence for the warm and cold stages. Unfortunately, because of their particular chemical composition, it is very difficult to date ocean floor cores using the radioactive decay clocks that work well on land deposits, such as in radiocarbon, uranium-series or argon-argon dating.

ABOVE: The *Glomar Challenger* was the main ship of the Deep Sea Drilling project, from 1967 to 1983. In that time she drilled over 19,000 cores with a total length of 97,000 metres, covering 375,000 nautical miles of seas and oceans. Even more impressive were the data retrieved from these cores, which revolutionized our knowledge of the Earth's climatic history.

Nevertheless, in the 1960s it proved possible to date ancient coral terraces by uranium-series decay that had apparently been laid down in the high sea levels of the last interglacial, suggesting that Emiliani's MIS 5 was about 100,000 years old. Even more useful, there is one clear dating signal that is worldwide and recorded in the cores. The Earth has a liquid iron core and, as it spins, a natural dynamo is produced, giving the planet an electrical and magnetic field, with a North and South Pole. For reasons that are still poorly understood, but are thought to be related to instability in the liquid core, the Earth's magnetic field fluctuates, and periodically it completely switches from a situation like today – when a compass needle points to the north – to the opposite condition, where a compass needle would instead point to the south. These events are called palaeomagnetic reversals, and they have happened many times in the history of the Earth. When sediments are laid down, their metallic content can preserve a snapshot of the Earth's magnetic field at the time, and where the poles were. The last major reversal was about 780,000 years ago, and that switch is recorded in many deep-sea core records that were being laid down at the time. As we said earlier, Emiliani numbered the isotope stages, starting from the present-day warm stage as 1. The palaeomagnetic reversal occurs in the cores close to the transition between MIS 19–20, and in itself this is a revelation, since it means that while Penck and Bruckner and their followers recognized four cold stages and four warm, the deep-sea records have over twice as many in the last 800,000 years, and many more before that time. Given this advance in dating the cores, Shackleton, Imbrie and colleagues were able to show in the 1970s that Milankovitch's theoretical astronomical model and the data from the ocean cores fitted together extremely well – the Earth's orbital changes truly were the pacemakers of the ice ages. And this in turn made it possible to estimate the dates of the various marine stages much more accurately.

Despite the general progress that was being made in understanding the ice ages during the 1960s and 1970s, a fierce debate was developing in Britain at that time about how to build up an accurate local picture of the ice age sequence. On the one hand, there were those who relied on vegetation changes recorded from pollen in ancient lakebeds, such as the one at Hoxne. In this view, enshrined in a detailed 1973 report by the British Geological Society, there were four interglacials – the Cromerian, Hoxnian, Ipswichian and Holocene (our present one) – with three glaciations in between, the Anglian, Wolstonian and Devensian. On the other hand there were palaeontologists such as my late colleague at the Natural History Museum, Tony Sutcliffe, who believed that the sequence of fossil mammals could give a better picture of the changing climates of Britain than could the pollen record. For example, based on pollen data, the Ipswichian interglacial included sites that contained hippos and elephants on the one hand, and horses and mammoths on the other. The pollen people argued that this merely reflected change within the interglacial. It would have been relatively cold at its beginning and end (when horse and mammoth could have lived), and warm in the middle (with the hippos and elephants). Sutcliffe said that only the hippo-elephant fauna was truly Ipswichian, while the sites with horses and mammoths were in a different geological position, and represented an earlier interglacial, recorded at sites in Essex such as Ilford and Aveley.

There was no room in the pollen-based scheme for such an interglacial, but Sutcliffe argued that there was unrecognized complexity in the British record, and in reality it was more like the picture emerging from the marine

isotope records. AHOB members Andy Currant (Tony's successor as the Museum's expert on Pleistocene mammals), Danielle Schreve and Roger Jacobi have taken this approach even further by developing Mammal Assemblage-Zones (MAZ) that typify periods of time and have provided an entirely new framework of which AHOB is making great use.

The River Thames has played an important, if indirect, part in helping to settle some of the big questions about the sequence of British ice ages. This is because of the significant change of course in its early history, when it was pushed southwards to its present position by a huge ice cap, the largest to cover southern Britain. Once the Thames was in its new position, it started to accumulate masses of new sediments that have given us many clues about the ancient human occupation of Britain. And there is a key signal in the marine cores of a particularly pronounced cold stage (MIS 12) that occurred about 450,000 years ago, and which was followed by a particularly marked warm stage about 400,000 years ago (MIS 11). That cold stage almost certainly corresponds to the Anglian ice advance in Britain and the diversion of the Thames. The succeeding warm stage is then the Hoxnian interglacial, in which we can now place the famous Swanscombe skull, as we will see in the next chapter. But what about the warm period about 500,000 years ago, before the Anglian ice swept down most of Britain? This stage, MIS 13, was apparently not quite as warm or prolonged as the Hoxnian, and as we have seen it was thought until recently that people never made it to Britain at that time, and Swanscombe was our oldest evidence of human occupation.

Opinions started to change after 1969, when a site at Westbury in the Mendip Hills of Somerset began to be exposed during limestone quarrying. A small earthy fissure appeared in the top of the north face of the quarry after blasting, and fossil animal bones began to tumble out of it, including those of bears and rhinos. At first it was thought that the fossils indicated a late Pleistocene age, perhaps only about 100,000 years ago, but gradually some very unusual species were noted, ones that were rare or previously completely unknown in Britain. These included a primitive form of cave bear, a primitive rhinoceros, a dhole (wild dog), a jaguar, and a scimitar-toothed cat, suggesting that Westbury must be older than Hoxnian sites like Swanscombe. Yet the vole clock indicated that Westbury could not be as old as the Cromerian interglacial

OPPOSITE: Cores from ocean floors, lakebeds and ice caps reveal records of past climatic changes. This polarized light micrograph of an ice sample was extracted in Antarctica to study the atmospheric concentration of methane, a powerful greenhouse gas, by analysing the air pockets trapped in the ice core.

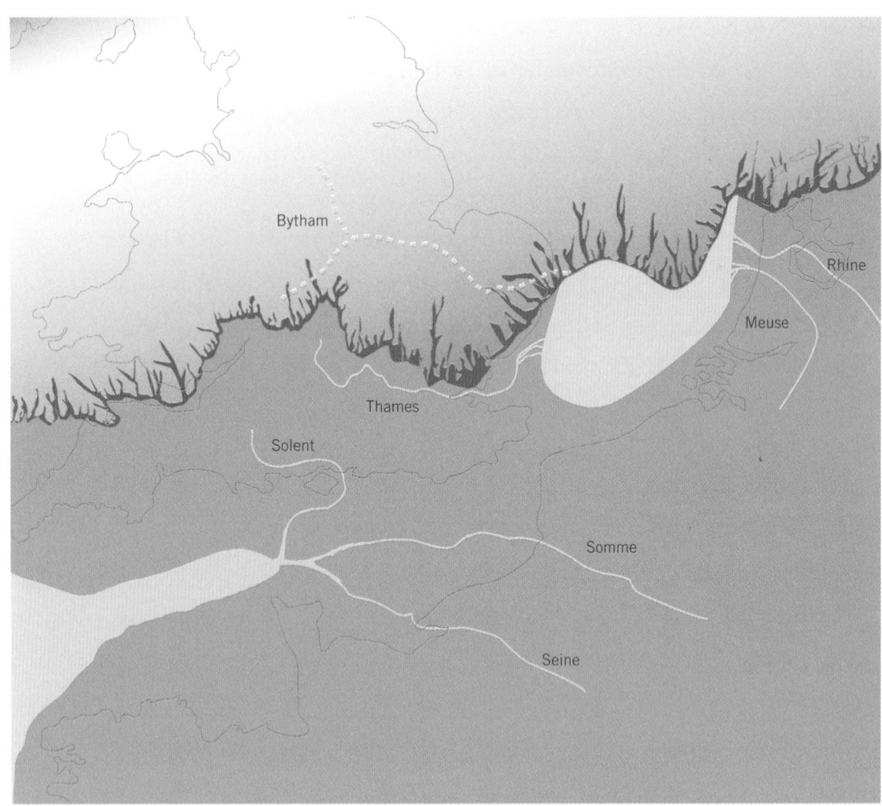

RIGHT: A reconstruction of Britain during the peak of the Anglian glaciation about 450,000 years ago. This destroyed the Bytham River and displaced the proto-Thames towards its present course. Immediately south of the ice, Britain would have resembled a polar desert.

deposits of East Anglia. As we have seen, the vole clock is based on the evolutionary transition between the primitive species *Mimomys savini* and its descendant *Arvicola terrestris cantiana.* Cromerian and pre-Cromerian sites in Britain and Europe have *Mimomys*, while Hoxnian and later sites have *Arvicola* fossils. Mike Bishop, then a curator of geology at University College London, studied the Westbury fossils for his doctoral thesis and argued that the combination of typical Cromerian mammals with *Arvicola* meant that they probably represented a hitherto unrecognized interglacial stage in the British Pleistocene, one that lay between the Cromerian (estimated at about 600,000 years old) and the Hoxnian (about 400,000 years). Moreover Bishop had been given, and had himself collected, what looked like stone tools from the site. If these really were stone tools, and they were the same age as the fossil mammal bones, they were older than any others yet dated in Britain. Bishop published these in the journal *Nature* in 1975, and their age was estimated at about 500,000 years, but scientific opinion was divided about whether they really were artefacts, and whether they really were as old as Bishop claimed.

Meanwhile something awful happened to the Westbury site. A conflict had been developing between the quarrying company and investigators like Bishop. The quarrying company had seen the earthy fissure enlarge with every blast of the rock face, from an initial width of about 25 metres in 1969 to about 70 metres in 1973. From their point of view, the loose deposits, while interesting, were an inconvenient and potentially dangerous obstruction to their need to quarry limestone. To Mike Bishop and an enthusiastic group of local supporters the site was of international significance and should be preserved, not quarried. The quarry operators took matters into their own hands in 1974 when they carried out a massive blasting operation, probably with the intention of completely removing the fossiliferous site. To their chagrin, instead of the intrusive deposits disintegrating to the quarry floor, the exposure grew even wider – to about 120 metres. It was now abundantly clear that the 'fissure' was in fact a huge cave chamber, filled with sediments to a depth of more than 30 metres, and the blasting had exposed a cross-section of its whole length.

In the bottom of the cave were yellow and white sands and silts, and near the top were red, brown and yellow earthy deposits containing most of the

LEFT: A sample of fossil vole remains from Westbury – these rich concentrations probably originated from piles of owl pellets.

OPPOSITE: The author recording a sample at Westbury in 1977.

OVERLEAF: Herds of buffalo cross a river in the Okavango delta of Botswana. Such masses of big game migrating across the Mendips or the southern coastal plains of Britain would have been familiar sights to the early inhabitants of Westbury and Boxgrove.

bones. Bishop was now *persona non grata* at the quarry, but the operators were aware that their work (and reputation) was potentially compromised. In 1975 they began negotiations with the Natural History Museum to undertake a full investigation of the cave, and in 1976 I and colleagues from the Museum began excavations that were to dominate our lives for the next eight years, and keep us busy on research and publications for a further fifteen. Blasting had removed much of the layered strata that Bishop had been studying, and parts of the cave were now dangerously unstable. We had to learn mountaineering techniques and at times excavated by dangling on ropes more than 30 metres above the quarry floor. The fact that the quarry itself was perched on the edge of the Mendips more than 200 metres above the Somerset levels, with distant vistas towards Glastonbury in one direction and the Bristol Channel in the other, sometimes made us feel we were parachuting or hang-gliding, particularly when strong westerly winds were blowing!

Our investigations showed that Bishop had got the overall picture of Westbury Cave right, even if our larger-scale excavations showed that he had got some of the details wrong. We were able to show that the cave deposits were much more complex than even he had thought, since there were actually two intersecting cave chambers, each with their own sequence. Each chamber consisted of many more strata than he had been able to recognize, covering a longer time period and showing much more variation through time. The lowest deposits were probably over 800,000 years old, with sparse remains of mammals, including the primitive vole *Mimomys*. As we excavated higher and higher up the cave we found evidence of environmental fluctuations in the mammals, with at least two warm stages, and a very cold stage at the top of the cave sequence at both ends, which probably represented the Anglian glaciation. The advanced vole *Arvicola* was definitely present in all the higher deposits, along with a rich collection of Cromerian fossil mammals, as Mike had claimed. We also recovered more possible tools made of flint and chert. Though none were as sophisticated as the handaxe tools from Swanscombe, most experts are now convinced that their shape is the work of humans rather than Nature. Moreover, from near the top of the sequence at the eastern end, we found even clearer evidence that humans had been at Westbury at least 500,000 years ago, in the form of cut marks made by stone tools on the leg bone of a red deer. So Mike

RIGHT: The shape of this remarkable scatter of flint debris at Boxgrove shows where someone squatted down and made a flint handaxe. The heavier flakes fell close to his or her legs, the smaller ones scattered further away – then the knapper got up and carried the handaxe off, leaving the profile of their legs preserved for 500,000 years.

Bishop's pioneering views were vindicated: humans really were in Britain during a warm stage just before the Anglian ice age. But it would take another discovery, from a site in West Sussex, to show us who these people were.

The Boxgrove site is located in a quarry near Chichester, and for many years flint handaxes like those known from Swanscombe were being found above marine sands that were preserved and quarried there. The marine sands were evidence that the sea once reached 10 kilometres (6 miles) further north of its present shoreline near Bognor Regis, at a time when the Isle of Wight was still joined to the mainland and a huge river, ancestral to the Solent, flowed by. The sea cut a massive cliff 30 kilometres (20 miles) long and 100 metres high (comparable to Beachy Head today) into the chalk hills of what is now the South Downs, and early investigators assumed from the height of the ancient sea and the excellent quality of the handaxes that the site was probably about the same age as Swanscombe. However, archaeological investigations there gradually developed into a huge project, involving over forty specialists and dozens of excavators, and radically changed perceptions of the site and its age.

ABOVE: A beautifully made flint handaxe from Boxgrove.

Long ago, topsoil and chalk from the ancient sea cliffs fell into a pebbly beach below, where they mixed with the bones of flatfish and conger eels feeding in the intertidal zone. The land was slowly rising and the sea level falling, and as the sea gradually retreated it left several metres' depth of fine sand behind while a huge lagoon, surrounded by salt marshes and grasslands, formed. Sand gave way to silts, and the next metre of deposits at Boxgrove is the most critical in what it preserves. Herds of game grazed on this new coastal plain – animals such as red deer, bison, horse, and even elephant and rhinoceros, as well as the animals that preyed on them, such as lion, hyaena and wolf. And people now became part of the local landscape, living off the land and the game. They were also drawn by the presence of flint in the chalk cliffs, an excellent source of raw material from which they could produce the characteristic stone tool found at Boxgrove – the handaxe – of which over 300 examples have been excavated. Because the land surfaces at Boxgrove were repeatedly covered over by gently flowing water, sealing them with a fine silt, those ancient surfaces have been protected with only minimal disturbance. The preservation is so good that the exact spots where people crouched down to make their stone tools have been conserved, with every flake of flint they struck off lying where it fell hundreds of thousands of years ago.

Not only that, but the bones of the animals they ate were also there, surrounded by tools, and often covered in butchery marks. The handaxes, which are predominantly oval or almond-shaped, were used to fillet carcasses of giant deer, red deer, bison, horse and rhinoceros. There are very few traces of cut marks on any of the bones from smaller animals such as roe deer, suggesting that smaller carcasses were either ignored or carried elsewhere for butchery. It is clear that the humans at Boxgrove had access to complete carcasses of big game, since most parts of the animals are represented at the butchery sites, and cut marks always precede those of the teeth of scavengers such as wolf or hyaena. This evidence was revolutionary when it emerged in the 1990s because it showed that these people were much more capable at getting meat than many archaeologists had believed: either they were regularly hunting big game up to the size of rhinos, or they were getting primary access to the carcasses of already-dead animals in competition with dangerous predators up to the size of lions, an equally impressive achievement. And while secondary access for scavenging and marrow extraction was probably enough to fuel the first human expansions from Africa, primary access to carcasses that still included the intestines and offal would have provided a higher quality and variety of food.

The site is dated from its mammal remains to an interglacial in the Middle Pleistocene, and the species represented are very similar to those from Westbury, particularly allowing for the fact that as an inland cave Westbury had a different local environment from the coastal plain at Boxgrove. Large animals represented as fossils include elephant, rhino, horse, bison, boar, five species of deer, ten species of large and small carnivore, and thirty small species including bats, shrews and moles. The type of water vole found at Boxgrove is once again *Arvicola terrestris cantiana*, as at Westbury and the Mauer sand pit that produced the famous jawbone of *Homo heidelbergensis* in 1907. As with Westbury, it seems that the sequence at Boxgrove is succeeded by indications of the severe cold of that critical marker the Anglian

BELOW: This magazine cover shows one of the excavation surfaces at Boxgrove. In the centre is the butchered pelvis of a rhinoceros shown on p. 98. It is surrounded by handaxes and further flint and bone debris.

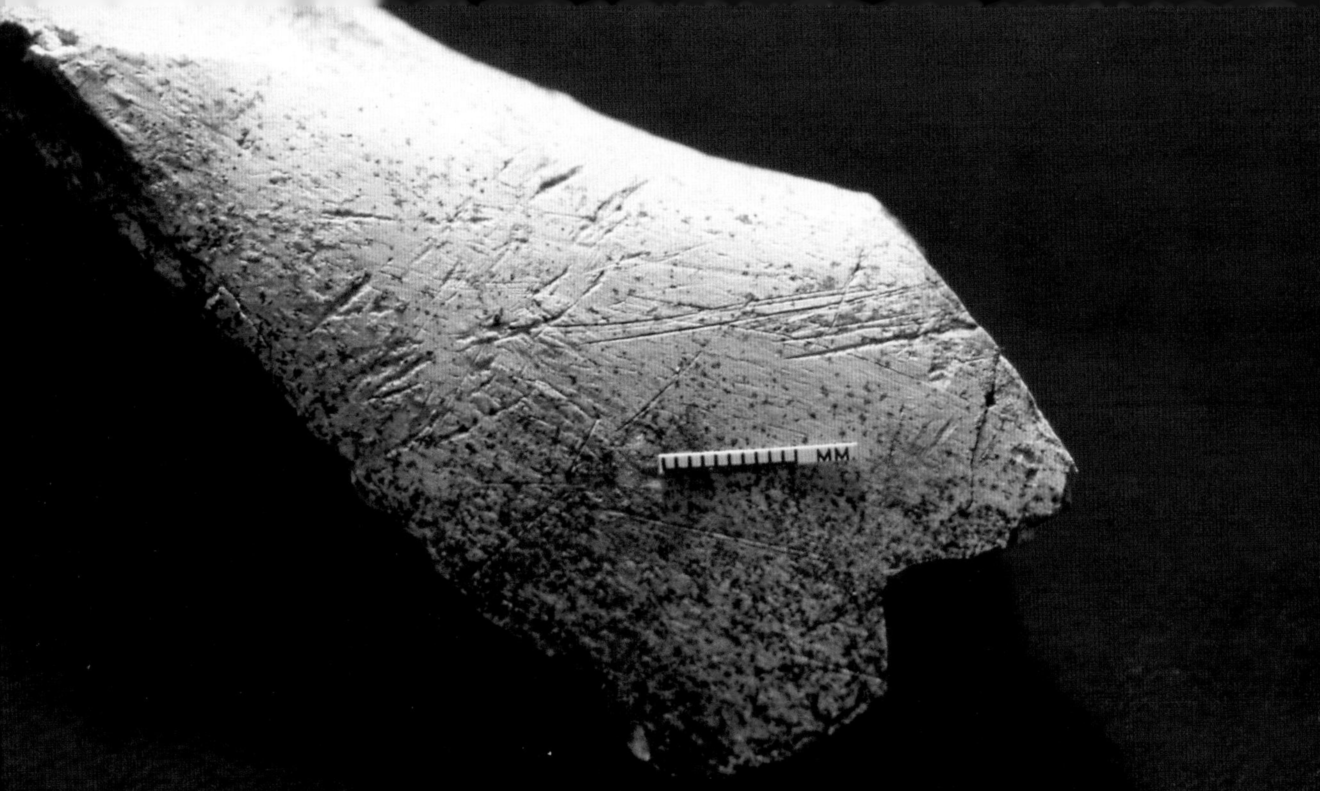

ABOVE: Cut marks made by flint tools on a horse bone from Boxgrove.

glaciation, suggesting that Boxgrove similarly belongs to the immediately preceding interglacial of Marine Isotope Stage 13, about 500,000 years old. The cold deposits look as if they have slumped down the slopes of the old chalk cliffs in spring thaws of snow and ice, and there is an intriguing hint that people did briefly try and cling to a precarious existence as the climate rapidly deteriorated, since the slurried soils contain a few stone tools.

It was not until the end of 1993, with a major discovery, that Boxgrove really hit the headlines. After nearly ten years of work, with his excavation funds running out, the Director of the Boxgrove Project Mark Roberts gambled his remaining budget on a last throw of the dice. He paid for a mechanical excavator and one volunteer, Roger Pedersen, to work over the winter months of 1993–4 on a previously unexcavated area of the site. Pedersen spotted a large bone in one of the trenches and immediately realized it was the limb bone of a large mammal – perhaps a human. A few days later Simon Parfitt and I confirmed that it was a human shin bone (tibia), and it was nicknamed Roger in honour of its discoverer. This was followed in 1995 by the excavation of two human teeth. Probably 100,000 years older than Swanscombe, these finds represent the earliest physical evidence of humans known from the British Isles, and the site has yielded a wealth of data on the behaviour of these people. We allocated the

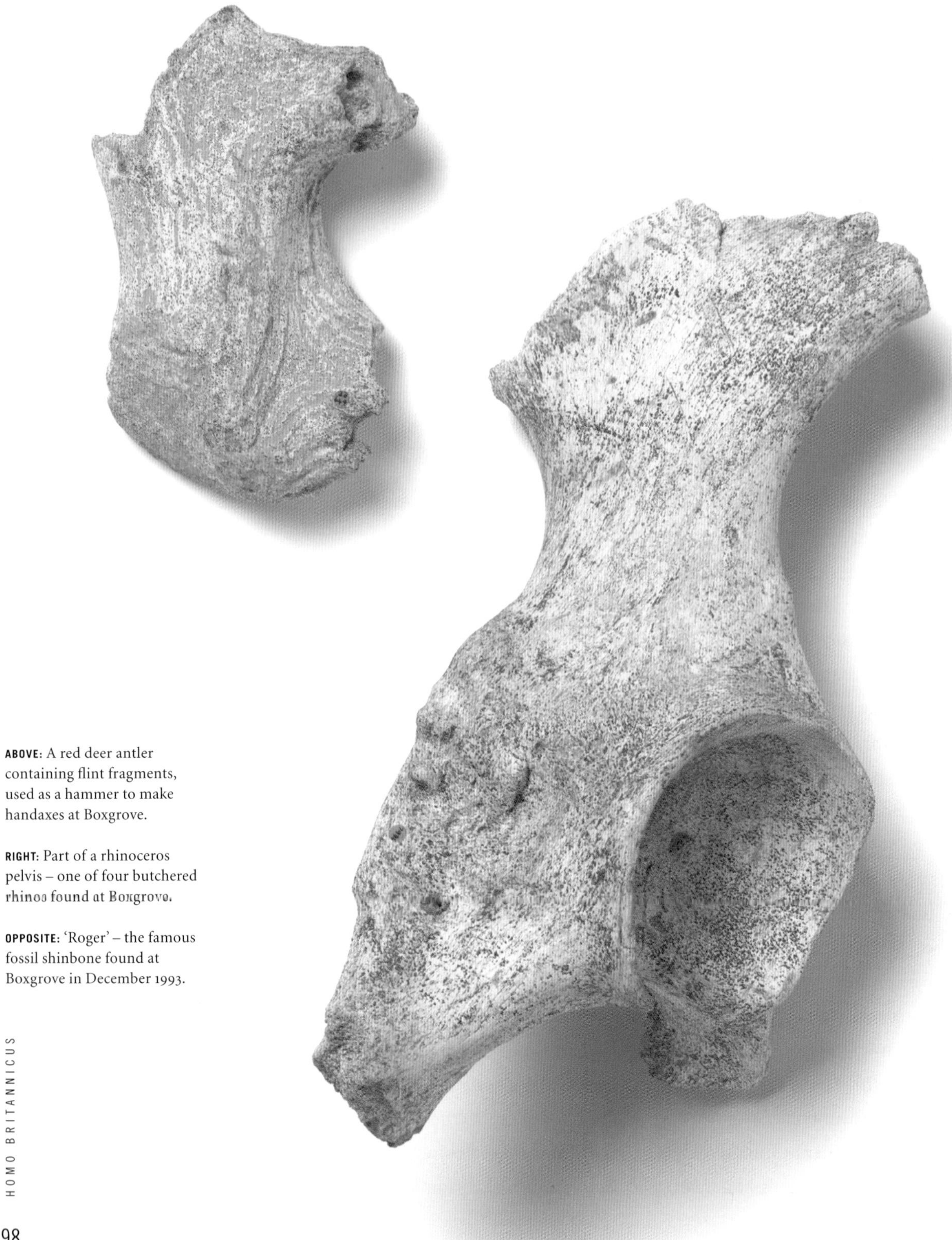

ABOVE: A red deer antler containing flint fragments, used as a hammer to make handaxes at Boxgrove.

RIGHT: Part of a rhinoceros pelvis – one of four butchered rhinos found at Boxgrove.

OPPOSITE: 'Roger' – the famous fossil shinbone found at Boxgrove in December 1993.

Boxgrove tibia to the same species – *Homo heidelbergensis* – as the Mauer jaw, and it is one of the most massive human leg bones ever found.

The individual concerned must have been about 1.8 metres (5ft 11 in) tall, and the great thickness of bone in the walls of the tibia suggests that this person (presumably a man, because of the size of the bone) was heavy and muscular, probably over 90 kilograms (200lb) in weight. The strength of the bone certainly reflects the physically demanding life-style that these people had to endure, but our microscopic study of its structure suggests that the individual had reached a good age at death – perhaps forty years old. There is no evidence on the bone to suggest the cause of death, although marks show that it was gnawed, probably by wolves.

The two human teeth found at Boxgrove are from the front of a lower jaw, and are not so exceptional in size, although heavily worn. Under a microscope, they reveal a mass of scratches and pits on their front surfaces. Many of these must have been made as the individual concerned sliced with stone tools through meat or vegetable materials clenched in their jaws. The direction of the slices can even be determined, and indicates that the tools concerned were being held in the right hand. The bases of the teeth are covered in tartar deposits, and these extend down the roots at the front, meaning that they must have been partly exposed during life, and that the teeth were probably being forced back and forth during chewing or clenching the jaws. The teeth probably fell out as the jawbone lay in the ground, and there is every chance that a British equivalent of the Mauer mandible still lies waiting to be discovered in the unexcavated remainder of the Boxgrove sediments.

The sheer antiquity of the Boxgrove human bones, at half a million years, is simply staggering, and caused a sensation in the media in 1994. One of the nicest illustrations of the time scale involved (and also of the chances of winning the National Lottery in Britain) came from a story and cartoon in the London *Evening Standard*. The cartoon showed Boxgrove Man in a block of ice, clutching a lottery ticket. The accompanying article made the point that if Boxgrove Man had been able to buy a ticket every week for the last 500,000 years he would be about due for a big win anytime now! While the odds of finding further human fossils at Boxgrove are certainly much better than that, what of finding even earlier human fossils at sites like Pakefield, or from the

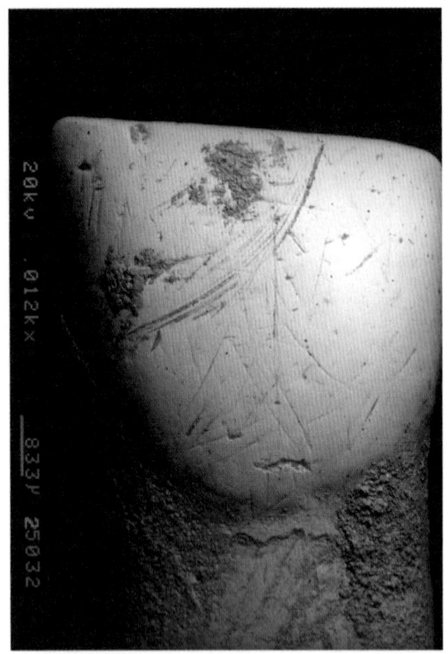

ABOVE: One of the human lower incisors found at Boxgrove. Both show scratch marks made by flint tools scraping across the front surface.

BELOW: From the *Evening Standard*: Boxgrove Man wins the lottery!

OPPOSITE: The Boxgrove site was excavated meticulously. Several handaxes can be seen in this view, which also shows the author's daughter Katy in the background.

time period between? Because they were able to trace fossiliferous beds over a large horizontal surface area at Boxgrove, the excavators increased their chances of finding important material, as well as revealing the associations between finds such as handaxes and animal bones on which butchery was carried out. By contrast, the deposits at Pakefield are at the bottom of a sea cliff, so it's impossible (or too dangerous) to excavate a wide exposure there. Somewhere like Norton Subcourse (in an inland quarry), or Happisburgh (on the beach), is much more promising in terms of the potential scale of excavation. But in one case, practical limits are set by the extent of quarrying, and in the other, by the fact that the deposits are under the sea most of the time. However, AHOB workers and their colleagues are constantly monitoring sites, and with the acceleration of erosion by sea and storms in parts of East Anglia, new exposures are appearing all the time. Maybe one day we will even have sites with human fossils to rival those from Gran Dolina and Ceprano.

It is now time to move on again, to 400,000 years ago, after the destructive forces of the Anglian ice advance had wiped out the descendants of the Boxgrove people, and forced the River Thames southwards to its present course, at Swanscombe in Kent.

Frozen deposits: Boxgrove Man might start to get lucky

CHAPTER THREE

THE GREAT INTERGLACIAL

PRECEDING PAGES: A line of African elephants march through grasslands. Even larger straight-tusked elephants trailed across Britain in each interglacial over the last 500,000 years.

For about a hundred millennia after the hunters quit Boxgrove, glacial Britain remained uninhabited. Northern Europe was similarly desolate, but further south at caves like Arago in southern France, and further east around the warm springs of Vértesszöllös in Hungary, humans survived. Then people returned, flourishing across southern Britain in a warm interglacial period. They were butchering their prey at Swanscombe in Kent, making wooden spears at Clacton-on-Sea in Essex, and building fires in woodland in what is now Suffolk. This long warm stage is known as the Hoxnian in Britain, but given the wealth of evidence of human occupation it perhaps still deserves its old name of the Great Interglacial.

In the Introduction, we saw how the stages of the Palaeolithic (Lower, Middle and Upper) were established. From the late nineteenth century onwards, archaeologists in Britain readily applied this system to sites that were already known when the terms were invented. Frere's handaxes were Lower Palaeolithic, much of the material that MacEnery and later Pengelly excavated from Kent's Cavern was Middle Palaeolithic, while Buckland's discovery at Paviland Cave was Upper Palaeolithic. As we have also seen, British geologists searched for the local equivalents of Penck and Bruckner's four glacial stages and three intervening interglacial stages. Gradually a classification was developed that partially, at least, matched the Alpine sequence. In Britain three separate glaciations could be recognized in the Pleistocene by the characteristic deposits they had left behind, and three alternating warm stages could be recognized from their characteristic plants and animals. The oldest were identified in East Anglia where warm-climate animals were fossilized under cliffs of glacial debris on the Norfolk and Suffolk coasts. The early warm stage was called the Cromerian, after the town of Cromer, and the succeeding cold stage the Anglian, after the regional name derived from the local Anglo-Saxon tribe, the Angli. The succeeding interglacial stage – the Hoxnian – was taken from John Frere's site of Hoxne where the pollen of warm climate plants was preserved in lake deposits, and the next cold stage was the Wolstonian, named after a Midlands locality containing cold stage deposits. The final two stages were the Ipswichian, named from a site in Ipswich where warm-climate pollen was preserved, and the Devensian, named after an ancient British tribe who lived in the vicinity of the type locality of cold deposits at Four Ashes in Staffordshire. It was thought that the evidence of the oldest Alpine glaciation – the Gunz – was either missing or not yet recognized in Britain, but the Anglian was probably the equivalent of the Mindel, the Wolstonian of the Riss, and the Devensian of the Würm. The Great Interglacial was the Mindel-Riss warm stage in the Alps, and the supposedly equivalent Hoxnian interglacial in Britain was when the handaxe makers seemed to reach their acme in skill and profusion. It was also the time period that at last delivered, in Kent, the long-sought remains of a really ancient Briton, at Swanscombe.

I explained in the last chapter how the Thames was pushed south by the Anglian ice advance, leaving behind abandoned valleys and gravel deposits.

MIDDLE TERRACE
HOLOCENE CHANNEL INFILLING
FLOODPLAIN TERRACE
FLOODPLAIN TERRACE

The river now formed a new series of deposits creating a staircase of terraces rising up on either side of its new route, and much of London and its eastern and western suburbs are built on these platforms. As we go up the staircase, we move back in time, and the highest of the steps is about 100 foot (30 metres) above present mean sea level. This highest step corresponds to the period immediately following the 450,000-year diversion, and it first contains glacial outwash from the Anglian ice cap, and then the deposits of the following interglacial, which we know in Britain as the Hoxnian. At that time the lower reaches of the ancient River Thames were several miles wide and it ran south of its present course at Dartford in Kent, laying down thick deposits that ranged from mud to sand to coarse stony gravels. These sediments were economically important in the nineteenth century and were quarried on an increasingly vast scale near the village of Swanscombe. The Swanscombe Cement Works was the first to perfect the production of Portland cement, and became its largest supplier, exporting the material right across the Empire. The quarrymen regularly came across bones of exotic animals such as elephant, rhinoceros and giant ox, and stone tools like those found by John Frere a century earlier at Hoxne – beautiful handaxes made from shiny brown flint. Harry Lewis, a Camberwell shoemaker, collected some of the first Swanscombe pieces and Victorian antiquarians started collecting material on a large scale from the 1870s; some estimates put the total number of handaxes found in the Swanscombe area at over 100,000. But in 1935 a find was made in one of the quarries, Barnfield Pit, that would elevate Swanscombe to fame as not only one of the richest of British Palaeolithic sites but also (as Piltdown lost credibility) the true home of the oldest English man (or English woman).

A dentist called Alvin Marston first collected stone tools and fossils at Swanscombe in 1933, and on the afternoon of Saturday 29 June 1935 he was alone in Barnfield Pit, apart from a few engineers overhauling the crushing plant. 'I noticed what appeared to be a piece of bone showing in the face about 6 foot above the floor . . . The object was cleared with the finger and then seen to be a human occipital bone.' Marston had to decide whether to collect it or leave it where it was, but as he did not have a camera with him he realized he would need an eyewitness of the bone's position in the ancient sediments. He was fortunate to find Frank Austen, the pit engineer, still working on the

OPPOSITE: The River Thames has accumulated vast and complex sediments in the London region over the last 400,000 years: the youngest and lowest fill the present river channel (Holocene Channel Infilling); those from the Floodplain Terrace include deposits from the last (Ipswichian) interglacial, and are about 120,000 years old; while the Middle Terrace includes material from about 200,000 years ago. This aerial view shows how the present river has truncated older Floodplain deposits and also shows (dotted area at the left) where the Trafalgar Square Floodplain material containing hippo bones has cut into older Middle Terrace sediments.

RIGHT: Alvin Marston indicating one of the findspots at Swanscombe (*left*), and with his son John (*right*).

crushing plant and, having sketched and measured the position of the bone, he took it 'to the nearest chemist, Mr S. Ackers, of Milton Street, Swanscombe, to be packed in cotton-wool'. Marston visited the site every weekend for the next nine months to continue searching the gravels, often sleeping there in a wooden hut with his teenage son John. 'The left parietal bone came to light on Sunday, 15 March, 1936. My boy and I were working. We had encountered a pocket of bones, including a rhinoceros tooth, earlier in the day, but in the late afternoon I began to uncover the parietal bone. There could be no doubt about it.' This time Marston left his son to guard the find and hurried off to ask Mr Ackers to photograph it *in situ*. Despite further careful searching over the next three years no more human fossils were found, and by the time World War II had broken out gravel extraction had stopped and the quarry face had slumped and was grown over with vegetation. Apparently, further gravel was removed without supervision in 1944 to help construct Mulberry harbours for the D-Day landings.

In the 1950s, another father and son team (B. and J. Wymer) were involved in locating and excavating the only surviving seam of the gravels that had contained the first two skull bones, as John Wymer explained:

> Work commenced on Friday, 29 July, 1955, with the assistance of Mr A. Gibson and my parents. It was while clearing this trial section, that, on the next day a human right parietal bone was found ... [it] lay dome upwards in the sandy gravel ... with a slight graze caused by the spade ... The bone was covered by an improvised shelter of tins and Mr A. T. Marston was contacted.

Unfortunately the bone had the consistency of wet soap, and when they tried lifting it using plaster, it broke into nine pieces, which were individually wrapped in wet tissue and newspaper and dispatched to join the other bones at the British Museum (Natural History). The Wymers continued excavating for the rest of the skull for another five years, but their search was in vain, and it may well be that it was originally there, but had been removed by quarrying before 1935, or in the wartime extraction of 1944. In which case it was powdered in concrete and may now lie on the floor of the English Channel, the eventual resting place of the Mulberry harbours.

Nevertheless, the three bones of the Swanscombe skull, scattered in the gravel 25 metres apart and found over a twenty-year period, met perfectly along the jagged sutures by which they originally knitted together in the head of their owner. Despite the fact that the bones are quite thick, to judge from the slight muscle markings this was probably a young woman, with a modern-sized brain

ABOVE: This small handaxe was found close to the human fossils at Swanscombe.

LEFT: A historic gathering at Swanscombe in 1955. John Wymer points to the location of the newly discovered right parietal, while the gentlemen in the background stand where the other two skull bones were found in 1935 and 1936.

OVERLEAF: An internal view of the three skull bones from Swanscombe (*left to right*): the right parietal (found in 1955), the occipital (1935) and left parietal (1936).

The structure of the brain, as far as can be judged from its outside appearance, was fundamentally like ours

close to 1300 ml in volume. The inside surfaces of the skull are so well preserved that they show the impressions of the blood vessels around the brain and even the foldings of the brain surface, making the specimen one of the most informative fossils in this respect, and suggesting that the structure of the brain, as far as can be judged from its outside appearance, was fundamentally like ours. Because the inner surfaces of the skull bones are better preserved than the weathered outer surfaces, this also suggests that the skull must have held together for some time in the river deposits before it came apart. But the fact that what was found represented only the back of the braincase meant there was a frustrating lack of the frontal bone, frustrating because the critical structure of the brow ridges, massive in Java Man and Neanderthal Man, and small or non-existent in modern humans (and Piltdown), remained unknown.

Thus the evolutionary position of the earliest English woman was in dispute. Some scientists, such as Sir Arthur Keith and Marcellin Boule, argued that Swanscombe would have had a domed forehead and a small brow ridge, like that of modern people. To them, Swanscombe was evidence that a 'presapiens' line, leading directly to modern humans, and parallel to the more primitive branch of the Neanderthals, existed in Britain during the Great Interglacial. In 1938, to back up his theory, Keith even published an exhaustive comparison of the Piltdown skull fragments with the Swanscombe parietal bone found two years earlier. Others, such as the German scientists Franz Weidenreich and Hans Weinert, argued that the shape of Swanscombe resembled the back of a fossil skull found at Steinheim in Germany in 1933. This skull was also thought to date from the Great Interglacial and, although somewhat smaller, had a frontal bone with enormous brow ridges. If they were right, it was possible that Steinheim and Swanscombe represented ancient ancestors of both the Neanderthals and modern humans.

More recent studies suggest yet another possibility. My doctoral research in 1974 compared measurements of the Swanscombe bones with those of modern humans, Neanderthals, and other fossils. In overall shape, Swanscombe was certainly not modern, with short, flat parietal bones, and a skull that was relatively broad across the base – in these respects it was most similar to Steinheim and to early Neanderthals such as a 125,000-year-old skull from Saccopastore in Italy. Such resemblances were confirmed when it was independently noted by the French researcher Jean-Jacques Hublin and the American researcher Albert Santa Luca that in the middle of the occipital bone Swanscombe had a small pit, called a suprainiac fossa. This pit, of uncertain function, lies at the upper edge of the attachment of the neck muscles, and while it is very rare in modern humans or human fossils, it is found in all known Neanderthals in which this part of the skull is preserved, and in Steinheim.

However, the strongest support for the Neanderthal affinities of Swanscombe comes not from Steinheim, but from discoveries a thousand miles to the south, near Burgos in northern Spain. During the 1800s a British mining company blasted a deep railway cutting through the rolling hills of the Sierra de Atapuerca, an eccentric act that would be frowned on today for its damage to the surrounding countryside. However, that railway cutting serendipitously opened up geological sections that have given us completely new windows into the world of the first Europeans. We have already looked through one of those windows, where 800,000-year-old fossil remains were found at Gran Dolina. But one cave in the hills had been providing a challenge to the young men of the region for centuries – they would impress their girl friends by descending into its depths and emerging with shiny cave bear teeth. The piles of cave sediment that they brought out accumulated near the entrance until in 1975 an expert on cave bears, drawn to the spot by news of the finds, began to inspect them. As well as the remains of cave bears, he spotted something very unexpected – a fossil human jawbone. This was the first find from what has become known, for very good reasons, as the Sima de los Huesos, the Pit of Bones. As I know from my own descent in 1992,

BELOW: Getting to the remote chamber of the Sima de los Huesos at Atapuerca requires this hazardous descent by ladder.

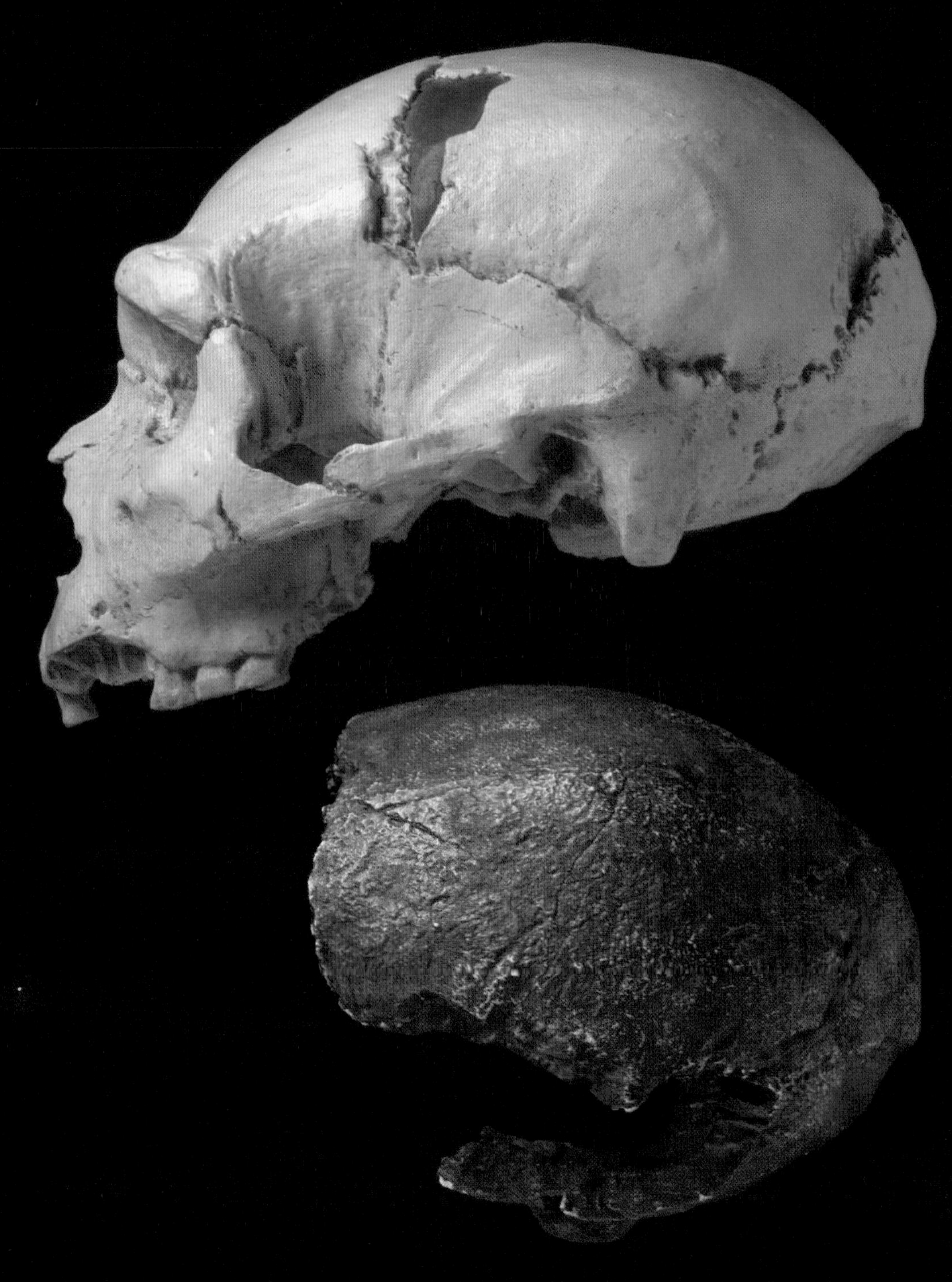

to get to the Sima takes over an hour of real caving, crawling, clambering, and using ropes. The final descent is made by a small chain-link ladder that dangles and spins into the darkness of a 15-metre deep hole. At the bottom of the ladder is an unprepossessing chamber so small and remote that visitors use up its oxygen in a few hours, and must vacate it regularly for the air to replenish itself.

OPPOSITE: A comparison of skull 5 from the Sima at Atapuerca (*top*) and Swanscombe (*bottom*). Both show incipient Neanderthal features.

When systematic excavations began in the 1980s it was found that cave bear bones were concentrated at one end of the chamber, and human bones at the other. After fifteen years of dedicated work under the most trying conditions (a steady temperature of 13°C, but a humidity of 100 per cent), the team working in the Sima have now recovered over 4,000 human bones and teeth from the Pit, remains representing about thirty men, women and children. Every part of the skeleton is there, but the sample is an unusual one. Judging by variation in the size and strength of the bones, and the stage of maturity they show, there are about equal numbers of males and females, but very few young and old; most were adolescents and young adults. Why there are so many, and how they ended up deep down a little chamber in a huge cave, are still unsolved mysteries. Some of the bones show signs of chewing by large carnivores such as lions, and this has led to speculation that the bodies of these people first lay elsewhere in the cave, perhaps the result of a natural catastrophe or an epidemic. They were chewed over by carnivores and later on were sludged in a jumbled mass by mudflows down into the Pit during wet phases of the cave's history. This would not only explain the chew marks but also why the bones of the skeleton are not represented in the proportions we would expect if complete bodies were decomposing in the Sima itself. However, the hypothesis put forward by most of the Atapuerca excavation team involves other humans throwing the bodies of their dead down into the Pit, in order to dispose of them. While there is no evidence that anyone ever lived in this part of the cave, there certainly was an ancient entrance, now blocked, much nearer to the Sima than the current opening on to the Atapuerca hills. And human involvement is suggested by the discovery of a well-made handaxe manufactured from a distinctive pinkish rock – the only artefact found associated with the bones in the Pit. Was this an offering to honour the dead?

However the bones found their way into the Pit, they give us a wonderful view of what the people of Europe were like at the time Swanscombe woman

ABOVE: Juan Luis Arsuaga (*left*), one of the Directors of the Atapuerca Project, with Stephen Aldhouse-Green, Director of the Pontnewydd excavations. Photographed by the author in the Sima de los Huesos.

OPPOSITE: A selection from the 2,500 fossil finds in the Sima de los Huesos.

was living alongside the ancient Thames. They were strongly built, both in terms of their muscularity (judged by the development of the pits and marks where muscles attached) and their bone thickness, especially in the legs. Many of the different bones can be reunited to create composite skeletons, and these show that a male individual known from a complete hip and leg bones was about 1.75 metres (5ft 9in) tall, and must have weighed over 95 kilograms (210lb). Females were nearly as tall, about 1.7 metres. Many of the large sample of teeth display markings that indicate growth disturbances in early childhood, perhaps from the time of coming off the breast and switching to solid foods. They are often heavily worn, but show no signs of decay, and we know from grooves between the back teeth that these people probably used toothpicks of wood or bone. The teeth themselves are somewhat larger than the average today, especially the front ones (incisors), and from wear and scratch marks on them we can see they must have often used their clenched incisors as an extra hand, to hold and cut or manipulate meat or vegetable matter. The bones generally give no clue about how these people died, although a number show signs of healed head injuries.

TIBIA

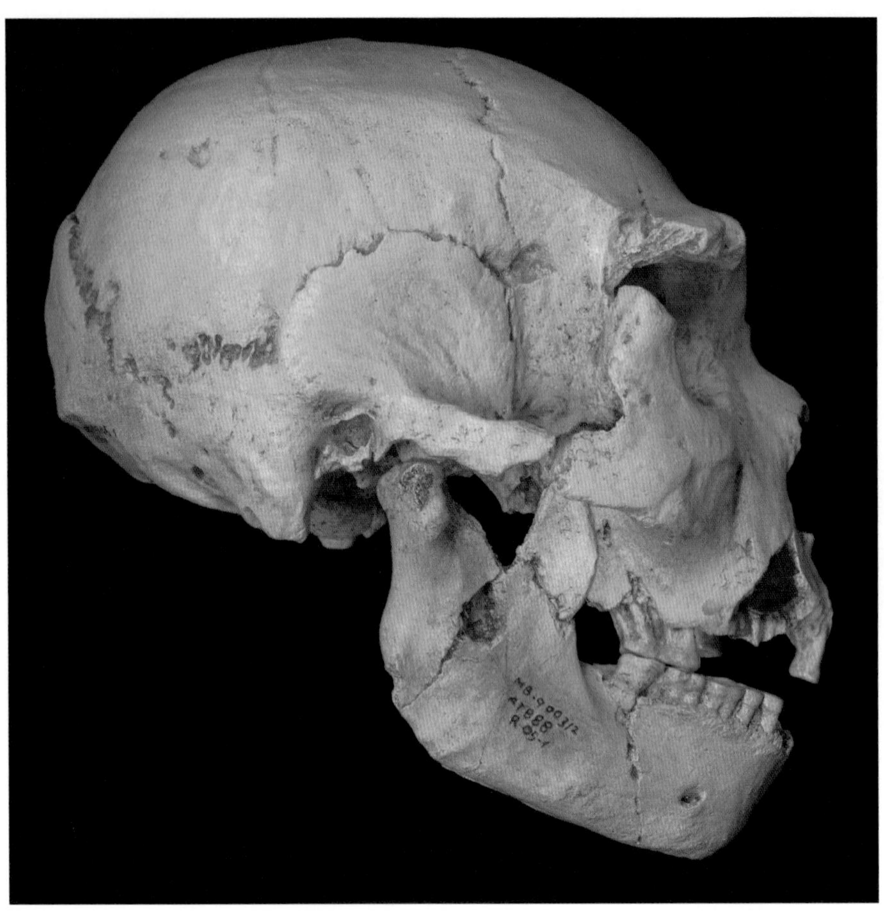

RIGHT: Skull 5, with its mandible. From tooth wear we know this is a rare example of an old individual in the Sima sample.

However, the best-preserved skull (number 5) is especially informative, with some thirteen possible injuries, and signs of serious inflammation of the face, which probably spread from abscesses in the teeth. In this case, the elderly man or woman probably died from the infection. A number of the skulls show signs of arthritis in the jaw joints, and one had diseased ears, and as a consequence may have been deaf. But overall, the skeletons are surprisingly healthy, with few other signs of injury or disease.

The heads of these people contained brains that, like Swanscombe, fall into the modern size range, although skull 5, at 1125 ml capacity, was very small. In modern people, the two hemispheres of the brain are asymmetric, and the shape differences can be correlated with handedness. The Atapuerca brain shape suggests that they, too, were predominantly right-handed, and this can be

The ears of the Atapuerca people were most receptive to the same frequencies as our own, perhaps an indication of the presence of speech

confirmed both by the greater size and muscularity of arm bones from the right side, and from the direction of the cut marks on the front teeth, where flint tools sometimes must have cut through whatever was being clamped in the jaws. The size of the brain gives little indication of its quality, but the preservation of the Atapuerca fossils is so good that there is another clue in the tiny bones of the middle ear – preserved in specimens such as skull 5. Using comparisons with modern data, it seems that the ears of the Atapuerca people were most receptive to the same frequencies as our own, perhaps an indication of the presence of speech.

In other respects the people of the Sima were very different from us. Their skulls were long and low, with jutting brow ridges over the eyes, and their jaws were strongly built, and chinless. At the back, the occipital bones show signs of the same little pit that we find on Swanscombe and on every Neanderthal – the suprainiac fossa. So it is likely that Swanscombe and the Atapuerca people were related to each other, and to the Neanderthals. We do not know what the face of the Swanscombe woman was like, although Steinheim gives us a clue, but the Sima samples show quite a range in shape. All the faces are large and broad, but some are rather flat, while others show a beaky nose, very like that of the Neanderthals.

When the Atapuerca fossils were first found, their age was estimated at about 250,000 years, similar to figures that used to be given for the Great Interglacial and for Swanscombe. However, we now think that both are actually about 400,000 years old. In the case of Atapuerca, this has come from uranium-decay dating of stalagmites which cover the Sima levels containing the human fossils. In the case of Swanscombe, the revision has come from new ways of dating the origin of the Thames staircase, as we discussed earlier. Although the Sima sample tells us so much more than Swanscombe or Steinheim can about the people of that time, the site itself contains virtually no evidence of the climate or possible adaptations of the people whose fossils are preserved there. For that we will turn back north again, to Britain and Germany.

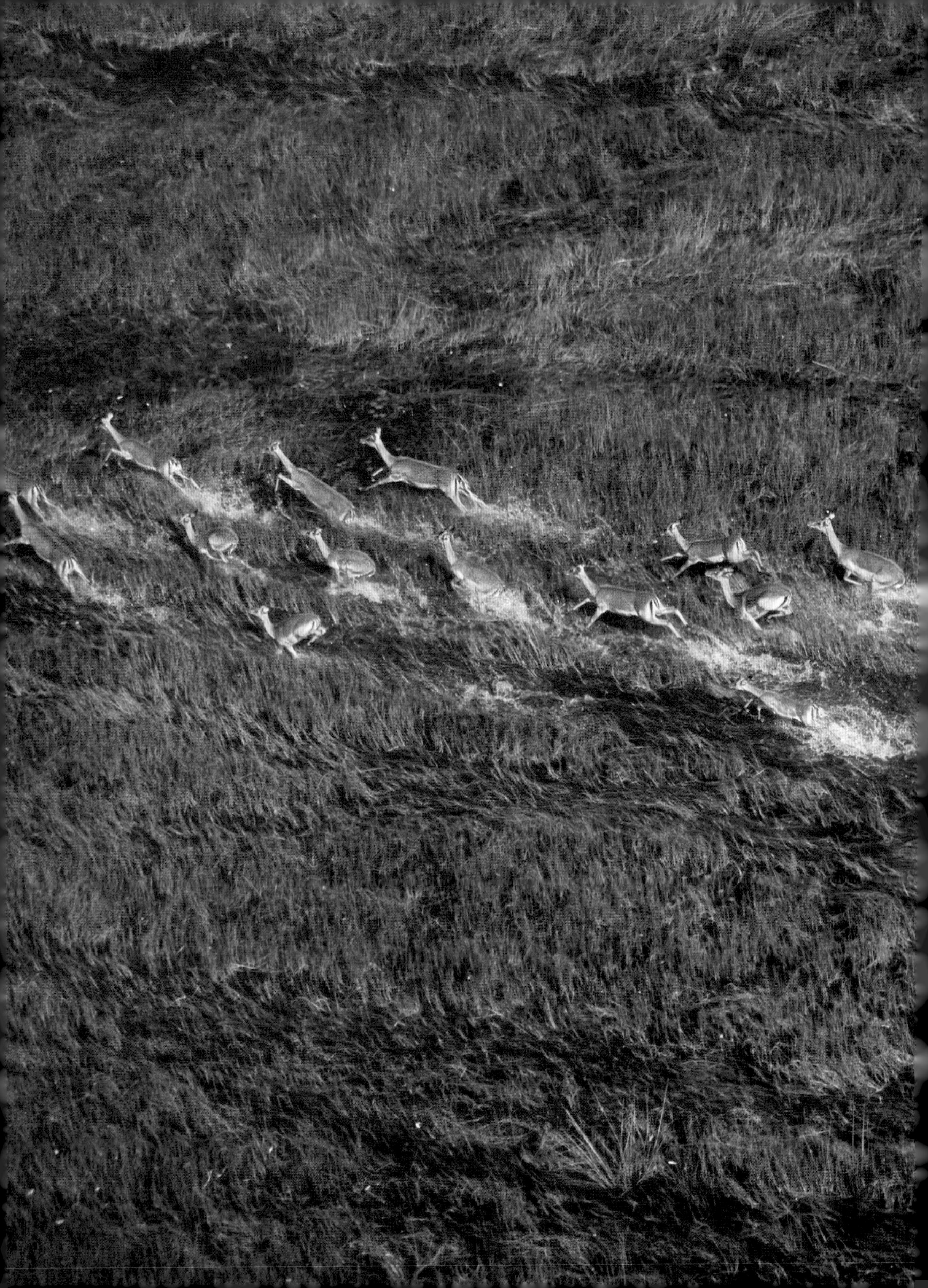

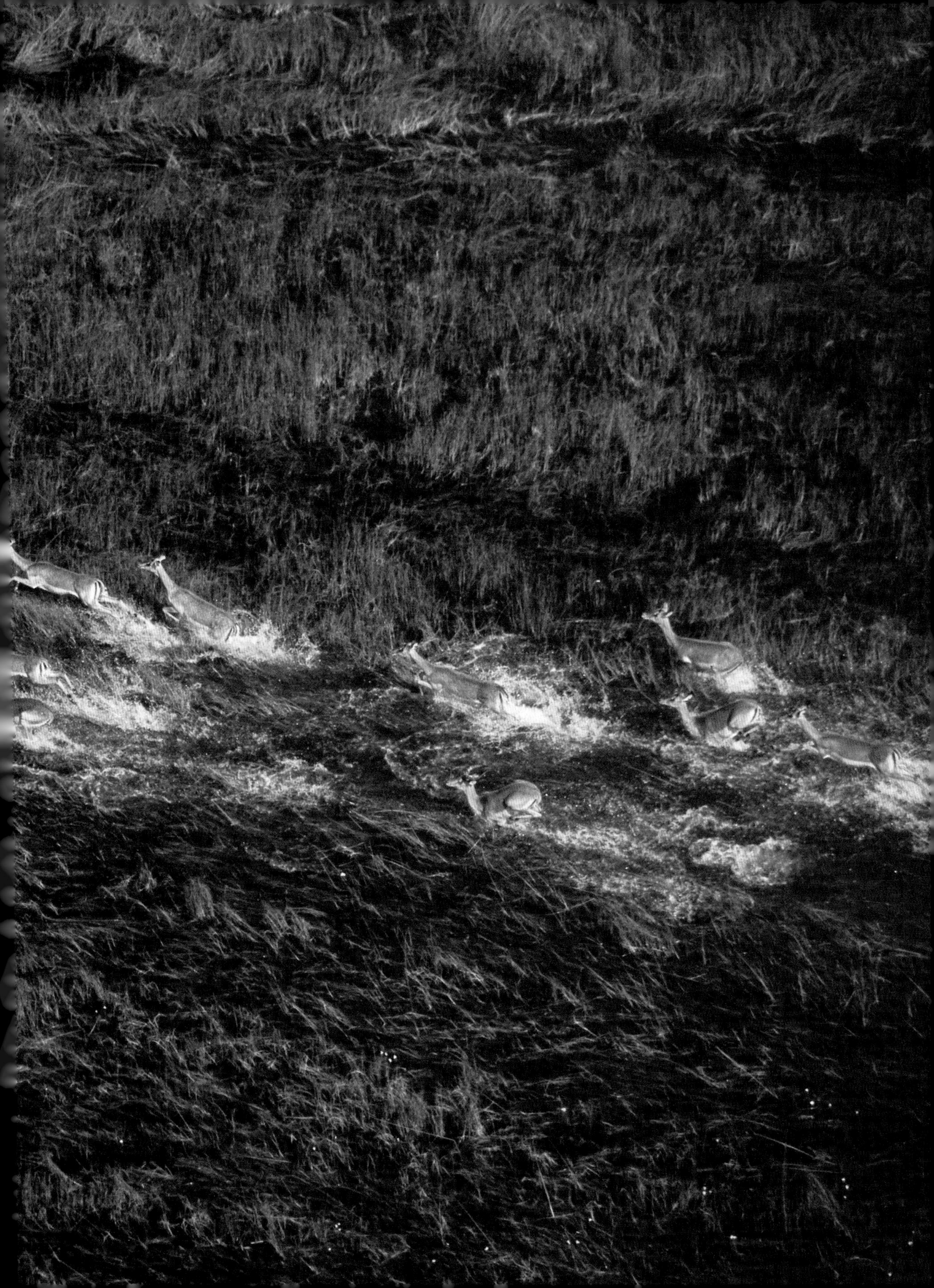

PRECEDING PAGES: Ancestral humans in Africa preyed off herds of antelope, like these. But the scene would not have been that different at Swanscombe 400,000 years ago, where three kinds of deer lived near the banks of the ancient Thames.

The landscape around Swanscombe when Swanscombe woman lived there would have looked familiar to us in many ways, once we strip off the vestiges of recent human 'civilization'. The ancient River Thames, miles wide, meandered over a lush floodplain, carving out and then abandoning many channels. In its waters lived freshwater mussels, and other mollusc species found today in the Danube and in rivers in India. There were also pike, eel, perch, salmon, dolphin and two types of beaver, and above the river hunting for fish were birds such as cormorants and osprey. In places there were marshy backwaters blanketed in reeds, in other places thick deposits of dried mud with grassland, inhabited by badger, marten, rabbit, hare, shrews and voles. Above the riverbanks were woodlands, with many of the trees we consider native to Britain today such as oak, alder and hazel. Emerging from the trees to drink or feed would be red deer, roe deer and fallow deer, although the fallow deer were unusually large and distinctively antlered. Animals that survived in the wilds of Britain a thousand

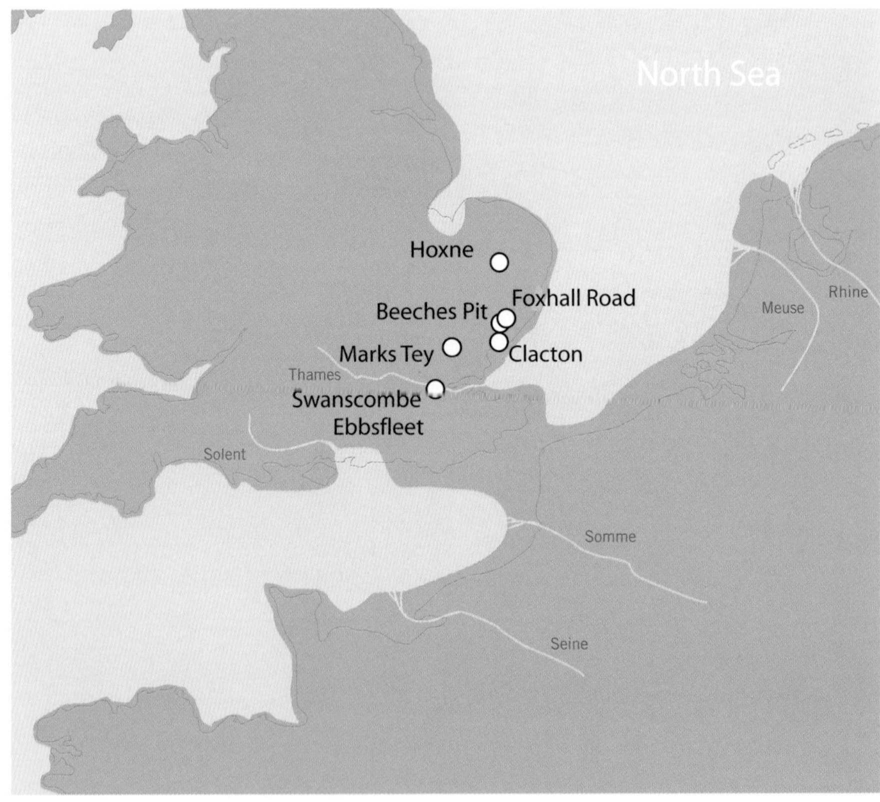

RIGHT: A reconstruction of Britain as it may have looked during the Hoxnian interglacial about 400,000 years ago. Some experts believe the English Channel had already formed by this time, but in this map the chalk landbridge to Europe is still intact.

years ago such as wild boar and wolves were also present, as well as herds of horse, bison, the now-extinct giant ox and giant deer. But added to these were much more exotic forms, some of which we might think would be more at home on the savannahs of East Africa than what is now the vicinity of the Dartford Tunnel. They included straight-tusked elephant, two kinds of rhinoceros, lions, and macaque monkeys. These mammals form one of the key assemblages of British interglacial mammals: the Swanscombe Mammal Assemblage-Zone.

Beneath the gravels that contained the Swanscombe skull and thousands of handaxes is a distinctive deposit of sand and silt called the Lower Loam that represents an earlier stage in the formation of the site. The Lower Loam seems to have been laid down in still or even stagnant water that dried out periodically, and at these times game and people walked across it, even leaving their footprints. Fossil remains of rhinoceros and the characteristic fallow deer were excavated, and the full antler development of the deer showed that they all died in the autumn or winter, as they shed and grow new antlers every spring. What is especially fascinating is that the people did not make handaxes. Instead they produced an industry based on flake tools known as the Clactonian, originally identified at the Essex site of Clacton-on-Sea, some 65 kilometres (40 miles) north of the present Thames estuary. This difference in behaviour has generated much discussion, including within AHOB. Were these essentially the same people as we find later at Swanscombe, but just making different kinds of tools for some reason, say for working wood rather than butchering animals? Or were they people who had an entirely separate way of living and working, who either did not know how to make handaxes, or chose not to, for reasons of tradition? And if they were there before the handaxe makers at Swanscombe, did they give rise to the later inhabitants, did they die out, or did they survive and live alongside them?

The climate and environment at the time of the Clactonian people at Swanscombe looks much like that prevailing when Swanscombe woman lived thousands of years later: warm, with grassland and trees and many of the same species of animal. But there are further clues to the nature of the Clactonian from a site about a mile away from Swanscombe, at Ebbsfleet. Large-scale excavations there for the Channel Tunnel Rail Link and associated engineering

E.
4740
Milton St.
Swanscombe
E.
1434

work since 1997 have been accompanied by archaeological investigations and, in 2003, just as the work was finishing, the skeleton of a straight-tusked elephant was found surrounded by about a hundred stone tools – Clactonian tools. As at Swanscombe, pollen and other evidence suggested that this Clactonian occupation was early in the warm part of the same interglacial period, in other words about the same time as the Swanscombe Clactonian. But the association of Clactonian tools and a large mammal skeleton was very significant because it had previously been suggested that handaxes were the preferred tools for butchery at this time, yet there are none at Ebbsfleet. So this evidence seems to support the idea of distinct populations with distinct traditions. That might also be indicated from the neighbouring and perhaps contemporaneous sites of Barnham and Elveden in Suffolk, where Clactonian tools and handaxes are preserved separately, but in comparable environments.

A chance discovery in 1911 at Clacton-on-Sea provides another insight into the way of life and capabilities of the Clactonian people. In waterlogged sediments that contained animal bones and Clactonian artefacts, a sharpened point made from yew wood had astonishingly been preserved. Microscopic study showed that it had been carefully shaped and had perhaps been hardened in a fire. The artefact was broken and less than a foot (30 cm) in length, which meant it could be interpreted as a spear point, a digging stick or even a snow probe (comparable to ones used in recent times by the Inuit in Alaska to search for frozen carcasses buried under snow).

The idea that it may represent a hitherto unrecognized, but important, component of Lower Palaeolithic technology has been strengthened by the discovery of wooden tools from a coal pit at Schöningen in Lower Saxony in Germany. This site, dated to between 350,000 and 400,000 years ago, has yielded eight partial or nearly complete wooden spears about 2 metres long, together with a shorter implement sharpened at both ends, perhaps a throwing stick, and smaller pieces of burnt wood. The spears were carefully worked from spruce and were found amongst some twenty horse skeletons, preserved by a unique mixture of acidity and alkalinity in the local soils. According to the German archaeologist Hartmut Thieme, who rescued them from destruction by mining machinery, they are carefully shaped and balanced like modern javelins, and thus were throwing spears. However, the American palaeoanthropologist Steve

PRECEDING PAGES: (*left*) Four handaxes from Swanscombe show the variety of sizes and shapes that were produced. (*right*) The Clacton spear: one of the oldest wooden artefacts in the world, made from yew.

RIGHT: Hartmut Thieme with one of the wooden spears excavated amongst horse skeletons at Schöningen.

Churchill has argued that the spears are simply too heavy and stout to be throwing spears, and anyway, without stone tips, they would have hardly pierced the thick hides of large herbivores. Instead, he suggests, they were thrusting spears that would have been used in close proximity to the prey. These are important considerations because killing at a distance (using javelins, slings, or bows and arrows) is much less dangerous and less physically demanding. Churchill suggests that this was a development of only the last 50,000 years or so of human evolution. My AHOB colleague Andy Currant thinks that the Clacton and Schöningen spears could have been used as goads to corner and wound an animal, and might also have been important in keeping the competition (other carnivores and scavengers) at bay.

Unfortunately, neither the Clacton nor the Schöningen site had any fossil human remains associated with the wooden spears, but a roughly contemporaneous German site near Weimar does preserve some skull pieces to compare with those found at Swanscombe and Atapuerca. The Steinrinne quarry near the village of Bilzingsleben has been mined as a source of travertines (water-deposited limestones) for several centuries. Fossil bones were found there in the sixteenth century and in 1818 Friedrich von Schlotheim found a skull encrusted in limestone, now sadly lost. Small stone tools began to be recognized in the travertine deposits, and in 1971 systematic excavations by the Halle Museum of Prehistory, and later by the University of Jena, began.

The travertine beautifully preserves the evidence of ancient vegetation, even delicate twigs and leaves, and we know that the local environment was warm and wet, with mixed oak woodland bordering a lake and springs. Animal fossils include fish, toads, snakes, birds and mammals such as straight-tusked elephant, rhinoceros, horse, three species of deer, lion, wolf, macaque monkey and giant beaver. Teeth and fragments of at least three human skulls were soon discovered and these have been reconstructed to show a long and primitive-looking braincase, with a very strong brow ridge at the front, and an angled rear. For some scientists, this is an example (perhaps the only one) of a European form of *Homo erectus*, but I think it is more likely to be related to European fossils such as those from Mauer, Boxgrove and the Greek cave of Petralona. There is no doubt that the skull bones do look more primitive than those from Swanscombe, and most of those from Atapuerca, even though the remains are all estimated to be around 400,000 years old. And while the skull level at Swanscombe produced thousands of handaxes, and the Pit of Bones at Atapuerca produced only one significant artefact – a handaxe – the sites

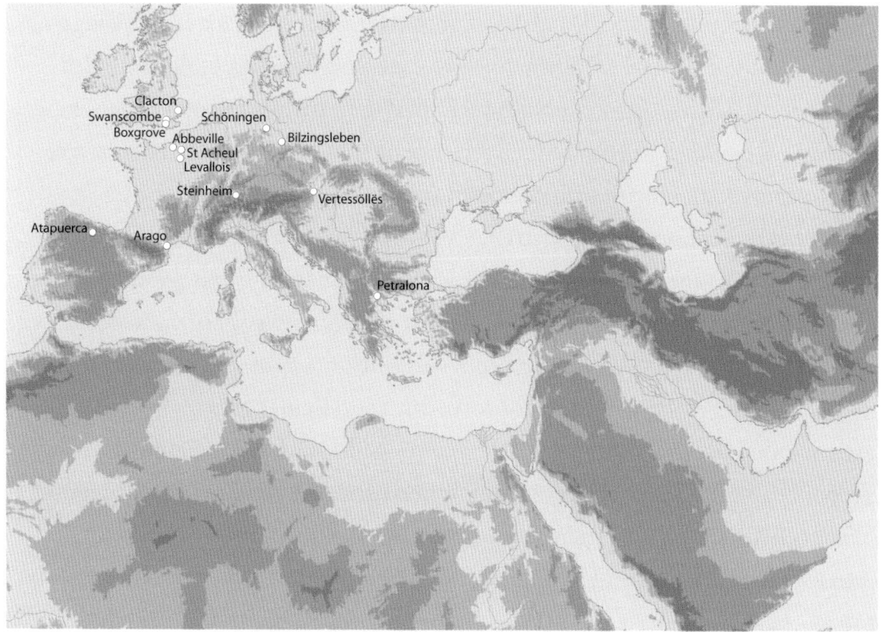

LEFT: A map showing some of the most important sites of this period of time in Europe.

ABOVE: A section showing the sequence of deposits at Swanscombe. The human skull bones were found in the Upper Middle Gravels.

of Clacton, Ebbsfleet, Swanscombe Lower Loam, Schöningen and Bilzingsleben all lack handaxes. It is tempting to suggest that ancestral Neanderthals like those known at Swanscombe and Atapuerca made handaxes, while the survivors of an earlier European lineage carried on flake tool industries such as the Clactonian and the tools found at Bilzingsleben. Intriguing as this is, it is speculating beyond the evidence we have so far. But given the continuing puzzle of the Clactonian, this is not an impossible scenario, and we await the discovery of a Clactonian fossil human in Britain.

Returning again to the handaxe makers of the Great Interglacial in Britain, unique insights have come from careful restudy by AHOB of finds from the Foxhall Road site, near Ipswich, excavated between 1903 and 1905. The stone tools were re-examined, along with the recently rediscovered archives of the dig, and the new study suggests that the site was initially a small lake or pond that later became part of a major river system. Unique clusters of distinctive handaxes, perhaps left around a small campfire, show that each handaxe maker had their own particular style. These differences could be due to their experience or their skill, but it is remarkable that the personal imprint of individuals from the Palaeolithic can be recognized some 400,000 years later.

And from another site of this age we seem to have the first good evidence of the use of fire in Britain. We do not know if the knowledge of controlling fire was imported or discovered here, but it certainly must have been a significant event. Fire itself provided protection, warmth and light, potentially extending the time that people could work, talk and socialize, and as a social focus it probably also fuelled social evolution. When used to cook, fire provided health benefits because cooking kills pathogens and parasites in meat, and it can deactivate toxins in plant foods. And it indirectly influenced physical evolution too, because by tenderizing meat and softening vegetables it removed the requirement to maintain large jaws and teeth, and the powerful muscles that worked them. The site in question is Beeches Pit near West Stow, Suffolk, which has been of scientific interest for over a century, and fortunately it has been protected as it lies within Thetford Forest. Here springs emerged through underlying chalk into woodlands in a warm and damp environment, laying down fine limestone deposits called tufa, similar to travertines. Tufa is a particularly good environment for preserving shells, and seventy-eight species of mollusc have so far been identified in a unique assemblage that has no modern equivalent. Some are completely extinct, others are found today in central or southern Europe, while one has its closest modern relatives in the Canary Islands. The presence of handaxes and the flint debris of their manufacture attests that people were here, and in the same levels there is plentiful charcoal, with concentrations that seem patterned – these look like the remnants of hearths rather than natural forest fires. In 1891 Thomas Huxley, Darwin's champion, described burnt bones of deer from the site, and recent excavations involving AHOB team members have found many further fragments of burnt bone.

ABOVE: Fossil shells of *Theodoxus serratiliniformis* found in Britain and Germany suggest that the Thames and Rhine river systems were linked during the Hoxnian.

Hoxne, the site of John Frere's discoveries of 1797, is still important today. Research and excavations in the last century showed that the sequence there actually began with deposits left from an ice advance. A 'kettle hole' formed in the deposits where a huge block of ice was left behind by retreating ice, and then melted. This large empty space became a lake, and the lake gradually accumulated clay in its bed. As they built up, these deposits also recorded the vegetation around the lake from the pollen of plants and trees that was washed in. Microscopic study of the pollen sequence allowed the changing local environment to be reconstructed, showing a whole cycle of warming and full forest development, through to cooling and the disappearance of the trees again as ice age conditions returned to Hoxne, and the lake became part of a river system. The sequence of vegetation changes was so well represented that the site was named as the type locality for the interglacial represented there, so the Great Interglacial is now known in Britain as the Hoxnian. The plant fossils from Hoxne showed that the lake was surrounded by oak forest. However, there was one puzzling aspect in that the middle of the warm stage was interrupted by an apparently colder, drier phase, with far fewer trees. Human presence was confirmed by the discovery of handaxes like Frere's at two different levels.

Because of the importance of Hoxne, AHOB has also excavated there, with rather surprising results. The work confirmed that, as expected, the basal deposits were glacial, followed by the warming recorded in the lakebed, leading to the formation of an alder carr (a dense and fertile wet woodland). Then the previously reported interruption occurred, shown by the intrusion of species such as dwarf willow and dwarf birch. As the deposits changed from lake to river, some warmth returned and it is only then in the sequence that artefacts were recovered by the AHOB team, at two related but distinct levels. So it seems that the human presence at Hoxne is only recorded near the top of the sediments, after the main Hoxnian interglacial sequence and not during it as had previously been believed. The handaxe makers occupied the region not when the climate was at its warmest, but later on, when it was somewhat cooler and the landscape was more open.

AHOB has re-excavated another site from this period at Marks Tey, in Essex, although this time the artefact evidence consists of just a few flakes and handaxes. Like Hoxne, the site is made up of deposits from a lake (over half a mile wide) formed in debris deposited by a previous ice advance. From pollen washed into the lake it preserves one of the most complete ancient vegetation sequences known in Britain. It began at the end of a cold stage with plants like sea buckthorn, proceeded into the following warm period with silver birch and pine, and then into a typical peak British interglacial mixed oak forest, including lime, ivy, holly, yew, mistletoe, hazel and alder. Gradually the climate and soils deteriorated as hornbeam, spruce and fir became more common, along with the exotic Caucasian wingnut, with birch and pine returning. Finally the cycle was completed with the return of grasses, wormwood and crowberry as cold again gripped the landscape. The preservation of finely layered deposits (rhythmites) that seem to reflect annual seasonal changes in water flow into the lake has even allowed the length of the warm stage to be calculated, at about 40,000 years. The site has been correlated with Hoxne because they both seem to immediately follow the same cold stage, and have very similar vegetation patterns (even down to the presence of a short, relatively treeless phase interrupting the warm period). This age has recently been confirmed through uranium-series dating

ABOVE: Some handaxes are not flaked all over, thus showing the original surface of the cobble or nodule from which they were made.

of carbonates laid down in the lake sediments, suggesting an antiquity of about 400,000 years. In keeping with a lake environment, the remains of fish, shellfish and freshwater microorganisms were recovered, and remnants of swan eggshell had even been found previously. The only bones of land animals found in the AHOB excavations were of deer and wood mouse, but on the opposite side of the A12 road is a former brickworks where remains of giant beaver, bear, horse, red deer, giant ox and elephant were reported by John Brown of Stanway as long ago as 1834. So this locality probably represents an environment closer to the ancient lakeshore at that time.

The Great Interglacial – the Hoxnian – provided one of the richest environments for early humans in Britain. There was lush vegetation and plenty of game to hunt or scavenge around a profusion of rivers and lakes. Early in the interglacial the Clactonian people were in eastern England, and later on there were handaxe makers. But the warm vegetation sequence was interrupted by a phase of deforestation and erosion, which is often associated in the deposits with evidence of burning. Had a large fire burned the forests, and could humans have been responsible? It has been suggested that they may have been practising an early form of a slash and burn economy, to thin out or even destroy dense

forests, and this may have backfired on them given the extent of environmental degradation it caused. But that does not seem to be the whole story, and the solution of this mystery is one focus of AHOB research.

Other AHOB research has thrown up some interesting patterns in human occupation during the Hoxnian interglacial. A number of archaeological sites began life as lakes on the landscape left by the retreating Anglian ice. Artefacts do not show straight away at these sites; they appear only when the deposits indicate a switch from lake to river. This suggests that in a heavily wooded environment, Hoxnian people were using the more open terrain along river systems as their thoroughfares through the landscape. These pathways would have had a variety of plants and animals and, in the river deposits, a ready source of rocks for stone toolmaking. That certainly seems to have been a major factor in why Swanscombe was such an attractive place for both Clactonians and handaxe makers.

The fact that sites like Swanscombe contain thousands of handaxes while others from the same period have virtually none also raises the question of what these tools were really being used for. In the case of the handaxe, this is one of the most discussed questions in Palaeolithic studies. These almond- or teardrop-shaped tools were made to consistent, often symmetrical, patterns, from many different kinds of rock, for over a million years. They were made by different species of early human, and can be found in sites spread across three continents: Africa, Europe, and Asia as far east as the Indian subcontinent. Flint-knappers of today can replicate these artefacts with about 10–20 minutes of concentrated effort, but it takes a long time to perfect the art, even for a modern human, and therefore some archaeologists have argued that the skill and consistency with which these tools were made must reflect the presence of a mind with our intelligence and complexity. And yet looked at another way, whether they are being made of volcanic rock to butcher an elephant beside a lake in East Africa a million years ago, or of flint to process a horse at Swanscombe 400,000 years ago, their variation in size and shape still seems limited, even rather monotonous. Thus to other archaeologists, the uniformity of the handaxe through time and space is peculiar. To them, it does not look like a product of a brain or behaviour like ours. When asked whether making handaxes so skilfully meant that early humans had language, the archaeologist Desmond Clark could only reply that if they *were* talking to each other, then they were saying the same thing over and over again!

In their search to understand the handaxe, archaeologists have replicated them, measured them, weighed them, thrown them, and studied them in microscopic detail to learn how they were made and, perhaps, what they were being used for. Until recently, it was generally believed that the first handaxes were poorly made but that, as human brainpower and skill improved, they became more and more refined. Thus the quality of an artefact could be used to date it – the cruder it was, the older it was, whereas if it approached an idealized type, it must have been an advanced and younger example. Recent studies, including several by AHOB members, have challenged the notion of an evolutionary sequence for handaxes, with the recognition that the local rocks available would have greatly influenced what kind of tool could be produced, and that handaxes were probably being made for a variety of reasons, and in different contexts and environments. Accordingly, even in one time and place there could have been a lot of variation in the kind of handaxes being made. It is evident from some sites that handaxes were made, and apparently discarded, in huge numbers, often looking as fresh and unused as the day they were made. Given the time and care used to make them, why were so many

Modern experiments on elephants that had died in zoos or in the wild show that handaxes are indeed fine butcher's tools

being manufactured and apparently abandoned, when simpler tools, even sharp flakes, might have been just as useful? And why, if the main intention was simply to produce a cutting or scraping edge, was it necessary to overdesign a handaxe with more shaping and symmetry than was required by function alone?

Although there have been quirkier takes on the function of handaxes, such as the suggestion that they were aerodynamically shaped projectiles used in hunting or defence, the popular view is that they were the Palaeolithic equivalent of the Swiss Army knife. They sat easily in the hand, they had a point at one end (or a chisel-like surface if broken across), a cutting or scraping edge down the side, and a thicker butt to use as a hammer. They were also a portable resource of raw material for making fresh flakes quickly, when these were needed. If they were well made, this was a testament to the skill of the manufacturer and his or her pride in their output. Microscopic studies of used handaxe edges suggest that they were employed for a variety of tasks including butchery, working wood and chopping plant materials, and modern experiments on elephants that had died in zoos or in the wild show that handaxes are indeed fine butcher's tools. But experiments also show that they do not have to be shaped anywhere near as perfectly as many Palaeolithic examples in order to do their job well.

An archaeologist, Steve Mithen, and a science writer, Marek Kohn, have recently argued that there are other dimensions that should be considered to understand the enigma of the handaxe, that we must also look at its possible social role. They propose that handaxes acted as signals between males and females as to who would make a good mate – so as well as being the Palaeolithic equivalent of the Swiss Army knife, they were also a kind of status symbol. In their model, from watching handaxe manufacture and use, young females could monitor which potential partners found the best raw material, produced the nicest handaxes, and procured the best output from their use in terms of food or perhaps producing further artefacts made of wood. The way in which the males cooperated or competed in the process would have provided a further signal of their social and mental abilities. The outcome of this mate selection would have been the perpetuation of the handaxe tradition through the relative genetic success of those best suited to carry it on. And in the process, many more, and many more 'perfect', handaxes would have been produced than were ever necessary for their purely utilitarian function.

Whatever their function, handaxes were long-lived in Britain, manufactured by *Homo heidelbergensis* at Boxgrove (and Happisburgh?) more than 500,000 years ago, by people at Swanscombe 400,000 years ago, and by their possible descendants over 300,000 years later. However, between the time of Swanscombe and the 60,000-year-old Neanderthal site of Lynford in Norfolk, we enter a portion of our ancient history that is as enigmatic and poorly known as the first appearance of people in Britain. Following the rich evidence from sites like Swanscombe, we might expect early humans to be flourishing in Britain. Instead, in this long period between 400,000 and 60,000 years ago, we seem to lose sight of them altogether for much of the time, and this puzzle is the subject of the next chapter.

OPPOSITE: A prepared core and a detached flake, typical of industries that appeared in Britain about 250,000 years ago.

BELOW: Stone tools covering 350,000 years of the British Palaeolithic: (*left*) a prepared core about 200,000 years old; (*centre*) a small Middle Palaeolithic (Neanderthal) handaxe; (*right*) a small handaxe from Swanscombe dating from about 400,000 years.

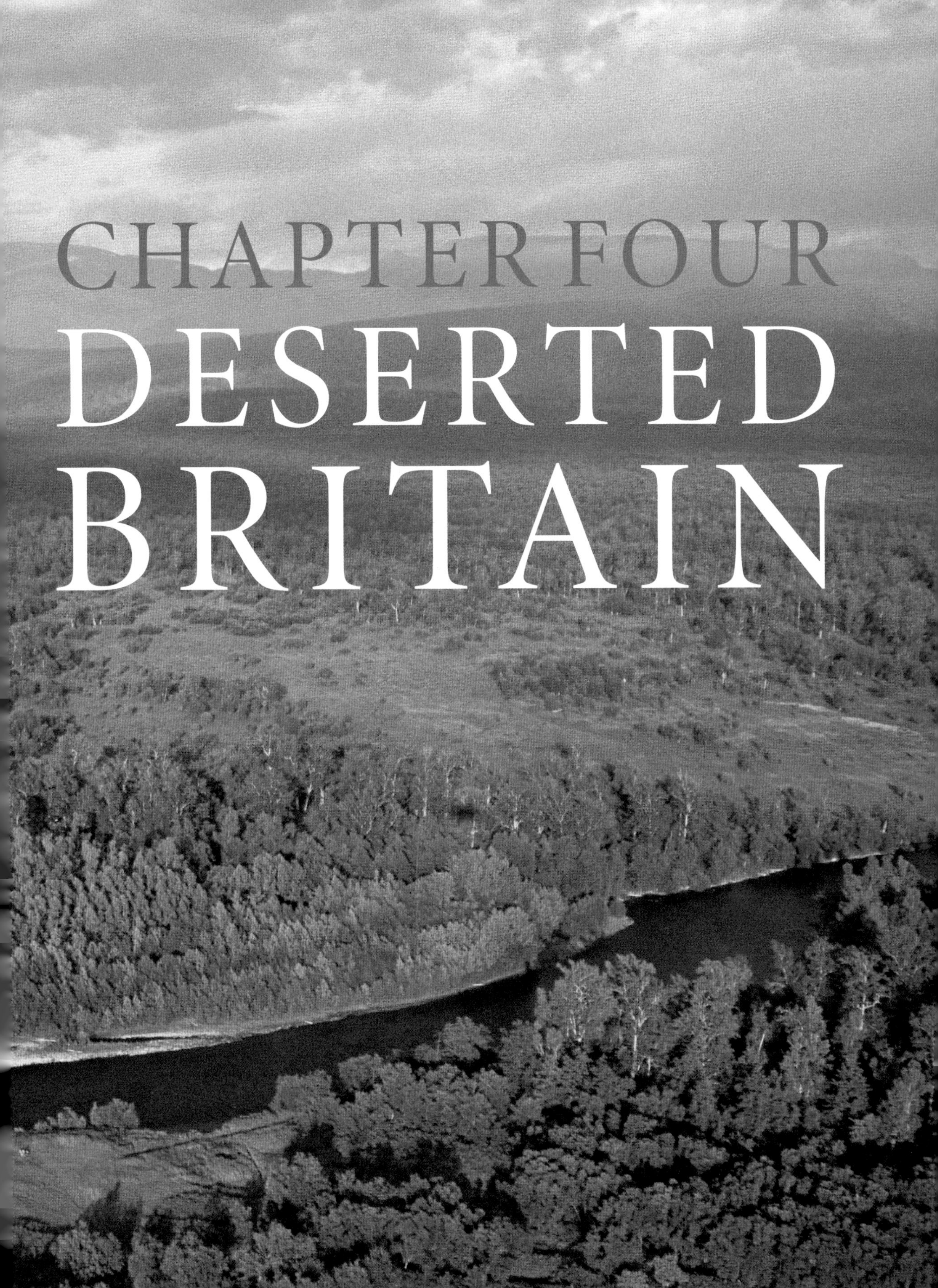

CHAPTER FOUR
DESERTED BRITAIN

PRECEDING PAGES:
The Kamchatka Peninsula, Russia. Heavily forested landscapes like this may have obstructed the early human colonization of Britain.

We saw in Chapter 1 how the first people reached Britain, then a small peninsula on the edge of the inhabited world, by 700,000 years ago. The environmental evidence from the Suffolk site of Pakefield is extraordinarily rich, but we can get only a shadowy glimpse of the early humans who lived there. That is not the case at the later sites of Boxgrove and Swanscombe, where there is extensive evidence not only of human activities but also actual remains of the people themselves. There are many sites like Swanscombe dating from the Hoxnian Interglacial, sites with thousands of beautifully made artefacts, and a strong signal of human presence. It looks as if the ancestors of the Neanderthals were making a good living from the local resources in the warm summers and mild winters of 400,000 years ago, and there seemed no obvious reason why they and their descendants should not have settled Britain permanently. And indeed it was a general scientific belief until recently that humans did persist in Britain from then on. It was thought that once humans had arrived in Britain, they would have maintained at least a foothold right through until the present day, over hundreds of thousands of years.

Even if one population replaced another, it was believed that process would have been essentially instantaneous, with no significant gap in occupation. But humans did not persist after the Great Interglacial. It came to an end, just as the preceding ones had, taking with it the lush river valleys, forests and grasslands on which the herds of horses and deer, and their hunters, relied. Ice sheets returned once again to the north-west of Europe, although this time they did not reach as far as the south of Britain. And a new pattern of episodic occupation was set in motion.

Archaeologists had long been aware that the record of human occupation in Britain was patchy, with some sites like Swanscombe and Warren Hill richly endowed with stone tools, while many others had just a handful. Judging human population numbers from counting artefacts is a hazardous business – for example, sites like Swanscombe and Warren Hill have been intensively scrutinized by collectors for over a hundred years while others, perhaps equally promising, have had much less attention. And another problem is whether a thousand artefacts in a riverbed represent the work of small groups living near the river over a period of a thousand years, or a large group working intensively over just a few days. Additionally, if the river flowed gently, small assemblages of tools could have been perfectly preserved, or if it surged fast, most of the artefacts were perhaps smashed or carried away. AHOB researchers have tried to address these problems by looking at archaeological records in the Middle Thames Valley. We saw in Chapter 3 that the Thames has deposited a staircase of terraces over the last 400,000 years, and because most of these terraces are in a densely populated region there are excellent records of their extent, depth and development. It is believed that each river terrace of sands, gravels and silts has taken about the same length of time to accumulate, and by looking at each terrace through its whole deposition, past variations in the flow of the river through time and space are averaged out. And because the Thames terraces have been quarried, built on and investigated by archaeologists for more than 150 years, tens of thousands of artefacts have been collected for study and cataloguing. Of course many, probably most, have been moved away from their original location, and the river also picks up and redeposits much older artefacts as it cuts into ancient gravels – so all other things being equal, the number of artefacts counted as the Thames staircase of terraces is descended would be expected to increase, particularly if humans improved their adaptations and thrived through time.

But that is not what AHOB researchers found. Instead, they noted that the number of stone tools collected by archaeologists in the sequence dropped, so over the time span of the Middle Thames terraces – about 400,000–100,000 years ago – there was a dramatically decreasing signal of human presence. In fact the signal virtually vanished after 200,000 years ago, with the few artefacts present in the terrace immediately above the present Thames floodplain arguably due to reworking from older terraces. What could this mean? The first thing to check was whether there was a bias in the samples – were some terraces bigger than others, were some more or less exposed, had some been more extensively quarried to yield a bigger sample to collectors? Many factors affected the rate and type of recovery. The majority of artefacts were collected (not excavated) by professional and amateur archaeologists who gathered them from active gravel pits or from the trenches of house foundations. This type of collecting was important from the 1850s until mechanized digging started to become common in the 1930s, as the urbanization of London took off and the pace of quarrying started to work against the observational skills of individual collectors. Another potential source of bias lay with the behaviour of the ancient humans themselves. As we shall see, there was a change in the types of tools being made over this period of time, and this meant that the archetypal (and for recent collectors highly desirable) handaxes of 400,000 years ago gave way to different-looking flake tools (which might not have been saved to the same extent by modern collectors). Additionally, if techniques of tool production changed and, for example, early humans needed to conserve their raw material to a greater extent, they may simply have discarded fewer of the tools they made. Nevertheless, even allowing for all these factors and biases, the pattern is certainly contrary to what might be expected. From about 400,000 years ago, humans seem to have had an increasingly hard time in Britain.

So let's see what evidence there is for human occupation in Britain between 400,000 and 70,000 years ago and then we'll examine what factors might lie behind the pattern. The glacial period that followed the rich sites of the Hoxnian, matching Marine Isotope Stage (MIS) 10, has no definite archaeology associated with it, but after that evidence reappears in Essex,

The presence of monkey may seem surprising, but it was still common in southern European woodlands at this time

on the opposite side of the Thames estuary from Swanscombe. At Purfleet, research since the 1960s in four now disused chalk quarries has given us a rare glimpse of human activity in Britain during the interglacial period represented by MIS 9, around 320,000 years ago. The deposits in question are some 8 metres thick and lie on ancient chalk bedrock, accumulated on one of the ancient Thames terraces when the river meandered north of its present course, and was actually running south-westwards at this location as it formed a huge reversed S-bend. Because of their relatively low altitude, the deposits were considered for many years to date from only the last interglacial, but fossils and archaeology both seemed to contradict this, and it has taken much careful work by geologists and palaeontologists, including members of the AHOB team, to establish their real age.

The sequence at Purfleet begins with a surface of frost-shattered chalk from the arctic conditions of MIS 10 about 350,000 years ago, followed by river gravels, sterile of fossils and artefacts. The next bed was laid down in a fast-flowing river with shells that indicate the climate was now warm, and the first artefacts appear, simple flakes and the cores from which they were struck. There are also remains of fallow deer, a species found in warm woodlands. The succeeding deposit is finely layered, suggesting tidal mudflats, followed by sands packed with freshwater shells, bones and occasional stone tools. These two beds clearly reflect a climate at least as warm as southern Britain today. There are fish such as carp and pike, lizards and snakes, and amphibians such as green frog. The small mammals include white-toothed shrew, water shrew and water vole, while large ones include fallow deer, roe deer, beaver, macaque monkey and straight-tusked elephant. The presence of monkey may seem surprising, but it was still common in southern European woodlands at this time. Tree and grass pollen suggest mixed oak forests with open areas, perhaps due to the river or to grazing or browsing by large mammals. Succeeding sands and gravels indicate strong river currents and contain some handaxe tools, but far fewer fossils. Finally, the river deposits show the arrival of more open conditions (there are

47E5
OLD KENT ROAD

TRAFALGAR SQ

PRECEDING PAGES: The Natural History Museum has shelves containing thousands of fossils from the Thames terraces in London. In this case, from left to right, a rhino humerus, an elephant vertebra, an aurochs (wild cattle) horn and skull.

fossil horses), and a new technology at the transition to the glacial conditions of MIS 8. This new technology is called Prepared Core or Levallois, after the suburb of Paris where it was first recognized in Palaeolithic tools over a century ago, and we will discuss its implications next.

Up until this time in the Palaeolithic, there were two foci in stone toolmaking. The core would be used to produce flakes, and both might be utilized. In the earliest stone tools, one or two flakes were struck off a pebble to produce a basic cutting edge and convert the pebble (the core) into a chopper (the tool). The flakes that had been struck off, with their simple sharp edges, might themselves also have been used as cutting tools. As the Palaeolithic progressed, handaxes began to be made in many parts of the western Old World, and here it seems that the toolmaker was most interested in the final form of the core, purposefully trimming off flakes on alternate sides of the core until the desired shape of handaxe was obtained. Although many flakes were also produced, they seem to have been less important. In Prepared Core (Levallois) technology, by contrast, the form of the flake to be produced dominated the process. First the edges of the cobble were knapped to give a rough oval shape, and then the upper surface of this core was trimmed. Next a large flake was removed across one end of the core to produce a flat striking platform. Finally, the end of the core was struck at right angles to the prepared platform, detaching a single flake off the core. The core itself is often called a tortoise core because of its resemblance, after the main flake had been detached, to a tortoise shell. There seem to have been two particular advantages to this technique. If the process worked successfully, the shape of the flake was predetermined, and a more usable cutting edge could potentially be produced from a core of given size. In addition, the long thin flake produced could easily be modified to produce a handaxe, a knife or a spear point.

The Prepared Core technique was so successful that it dominated toolmaking for the next 250,000 years in Europe, western Asia and Africa. Neanderthals and early modern humans used it throughout the Middle Palaeolithic, and because it forms a common heritage of these two groups it has led to a theory about its origins and their evolution. Two researchers in Cambridge, Robert Foley and Marta Lahr, have developed what is called the Mode 3 Hypothesis (Mode 3 being another name for Prepared Core/Middle

Palaeolithic industries). They have argued that the divergence between Neanderthals and *Homo sapiens* occurred around 250,000 years ago, following the development of Mode 3 (Prepared Core) technology. They relate this archaeological innovation to African populations of about 300,000 years ago, who invented the technique and then spread successfully across Africa, into western Asia, and then Europe. Those who stayed behind in Africa evolved into the first modern humans, while populations who spread into Europe replaced earlier humans (such as the Boxgrove and Swanscombe people) and evolved into the Neanderthals. However, there are serious problems with the Mode 3 Hypothesis. First, in Europe, there is little sign of an intrusive migration. Instead it seems much more likely that the roots of both the Neanderthals and their Prepared Core technique are local. As we saw in Chapter 3, the Swanscombe and Atapuerca Pit of Bones people date from about 400,000 years ago, and are associated with handaxe tools. Yet both show signs of Neanderthal features,

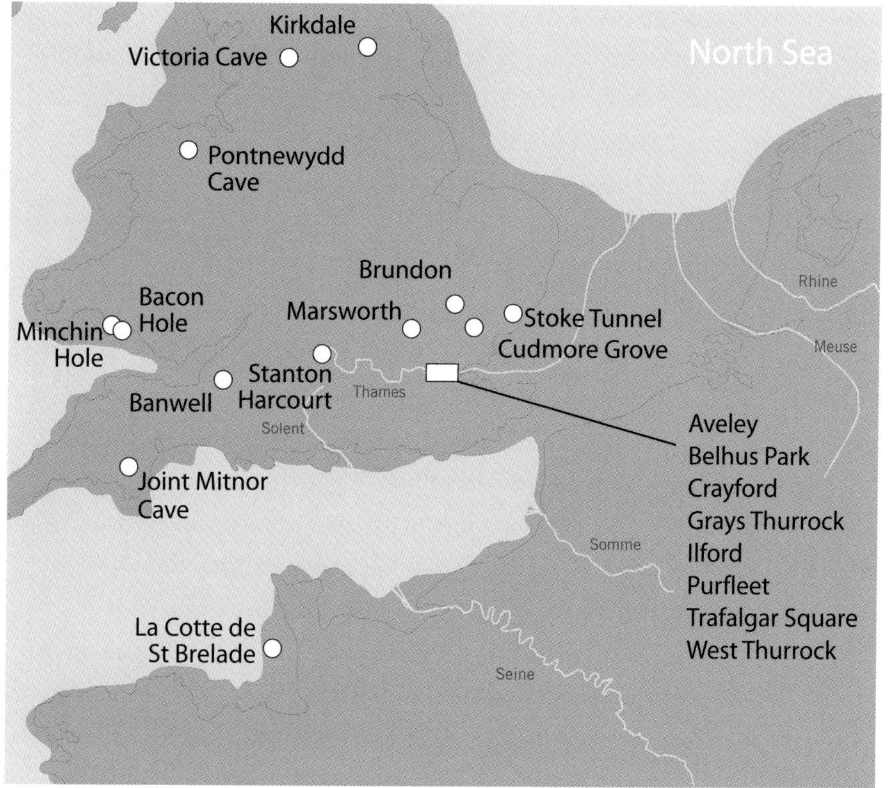

LEFT: A reconstruction of Britain as it may have looked during a time of intermediate sea level about 200,000 years ago (sites covering a wider time range are shown). Some experts believe that the English Channel had formed by this time, but the map reflects the view that it was formed later.

and can broadly be considered as Neanderthal ancestors, local to Europe and well before the date of the supposed migrations of tools and people ancestral to Neanderthals from Africa. Second, it is not clear where the Prepared Core technique originated, and it is possible that it spread through existing populations without replacement, or evolved in several regions independently, rather than spreading with a migrating wave of people.

Moving on to the mammals of Purfleet, these have been important in the recognition of the Purfleet Mammal Assemblage-Zone, which can be distinguished from the earlier Swanscombe Mammal Assemblage-Zone of 400,000 years ago by the presence or absence of certain diagnostic species. For example, an extinct small mole and giant beaver, rabbit, and European pine vole are all present in the earlier interglacial but not at Purfleet, while hyaena is known from Purfleet but not from the Swanscombe interglacial (the Hoxnian). Similarly Purfleet can be distinguished from later interglacials by species like the macaque, which is not known in Britain after this time. Moreover, similarities in mammals, pollen and other indicators suggest that further Essex sites such as Grays Thurrock and Belhus Park near the Thames, and Cudmore Grove near the River Blackwater, also belong to the Purfleet interglacial of about 320,000 years ago. Purfleet also has a special archaeological importance in its succession of artefacts, recording a sequence going from what looks like a Clactonian industry, without handaxes, through handaxe levels, and finally the appearance of Prepared Core artefacts. This suggests that the region remained important to humans through a considerable period and through marked changes in the climate and landscape.

The Purfleet archaeological record is significant because, in the apparent succession from Clactonian to handaxes, it duplicates what was recorded in the previous interglacial at sites like Swanscombe. We discussed the different ideas to explain the Clactonian/handaxe contrast in Chapter 3, and the recurrence of this pattern may support the idea that different populations were involved. If this is so, as Britain warmed up in both interglacials, the first wave of colonizers were the Clactonians, and people making handaxes followed them. We do not know where either group came from, but perhaps the Clactonians were geographically closer to Britain, more mobile, or more suited to the plants and animals of the early part of the interglacial. Whether the Clactonian people subsequently died out, migrated, or were replaced or absorbed by the handaxe makers, we do not know, and unfortunately we have no remains to show what they looked like in comparison with the Swanscombe fossil. In turn, during the Purfleet interglacial, the handaxe makers developed into, or were replaced by, people making Prepared Core tools. But these people were now struggling in the face of changing conditions, as the Purfleet interglacial drew to a close. Human visibility once again shrank to virtually nothing. One handaxe site, at Harnham near Salisbury, has recently been discovered and dated to the following cold stage, but otherwise we pick up the signal again about 100,000 years later, in a rather surprising place – North Wales.

The early Palaeolithic of Wales is hardly represented at all. While over 50,000 handaxes are known from England, there is only a handful from Wales, with one notable exception – Pontnewydd Cave. Part of the reason for the lack of tools in Wales must be the dominant highland terrain of the country, where repeated glaciations have destroyed the evidence, except in special cases such as Pontnewydd. Pontnewydd (New Bridge) Cave lies about ten kilometres inland,

BELOW: A plan view of Pontnewydd Cave. Most of the human fossils were recovered from the Main and East Passages, but one was excavated far away at the New Entrance.

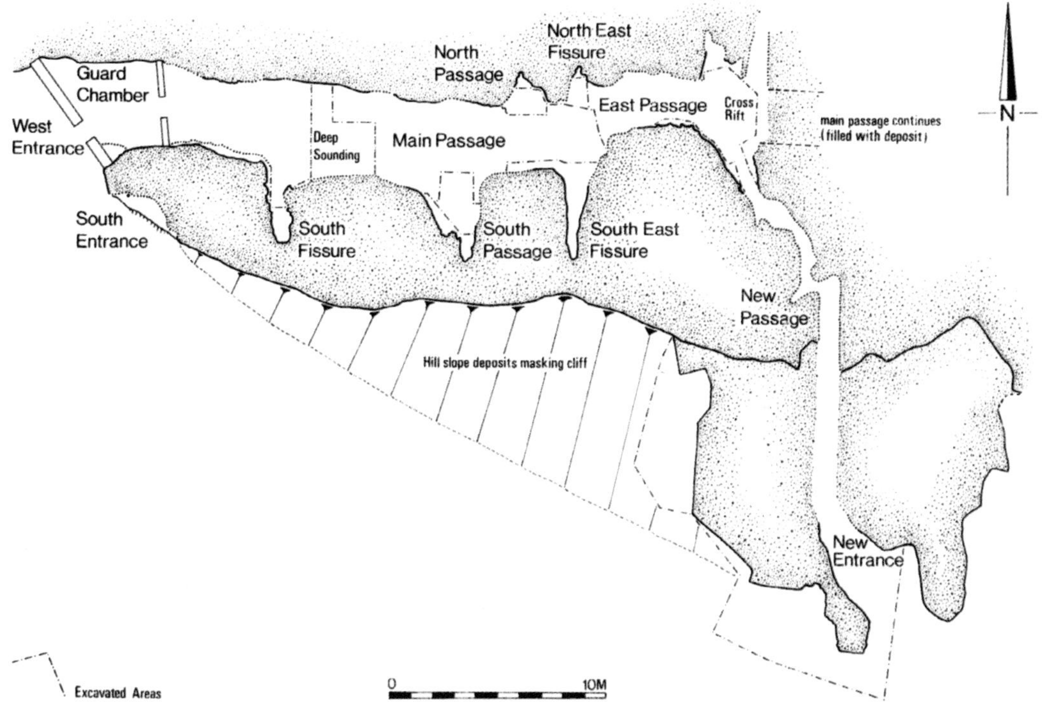

Much later still, long after the phase of human occupation, reindeer, musk ox and wolves roamed a treeless tundra

south of the seaside town of Rhyl and the broad Vale of Clwyd, in the lush Elwy Valley. The cave was mentioned by a Reverend Stanley in 1832 who had dug in the nearby Cefn Cave and he reported, 'I found [it] to be entirely blocked up with soil, and [it] has clearly never been open to human observation. But I have no doubt, from its appearance and character, that it will . . . exhibit as rich a prospect [as Cefn], whenever its recesses may be explored, in search of those organic remains now unknown in the temperate zones.' Forty years later Professor William Boyd Dawkins described his excavations with two local amateur prehistorians, which supposedly yielded fossil remains but no artefacts. However, later sorting of the large spoil heaps he left behind revealed many stone tools and animal bones that were missed in 1870. Soon afterwards, the geologist McKenny Hughes carried out more careful excavations and not only found fauna and artefacts but also a large human molar tooth. The anatomist George Busk, whom we met in the Introduction as someone who studied the Neanderthal skull from Gibraltar, examined it in 1874 and said it looked 'quite as ancient as the rest'. It was thought to date from the last glacial stage and was almost certainly an early Neanderthal tooth, the first ever found in Britain. But sadly it is now lost. Occasional small excavations continued at Pontnewydd until 1940, when the cave was turned into a wartime store for land mines and depth charges, with the addition of a concrete guard chamber and a grey wall for camouflage across the entrance.

In 1978 Stephen Aldhouse-Green (then of the National Museum of Wales) began new excavations at Pontnewydd that have revolutionized our view of the cave, its age and importance. They showed that the cave once had a series of occupations by early humans at the entrance, when rivers were flowing nearby, some 50 metres above the present river level. But ice sheets that had subsequently surged up the valley destroyed all trace of these ancient visits outside the cave. Fortunately, however, during these ice advances some of the cave entrance sediments became soggy and oozed down into the cave where they survived, not in their original form, but churned up during relocation.

After the deposits had been dumped in the cave, stalagmites grew over them, and these have been dated by the uranium-series method, showing that the oldest sludging must have happened, not in the last cold stage, but over 200,000 years ago. Any fossils or artefacts in the oldest mudflows therefore date from at least that time. Deep-sea records show that Marine Isotope Stage 7 was a warm period that began about 250,000 and ended about 200,000 years ago. However, it was a complex stage with a severe cold phase in the middle, around 230,000 years ago, and this provides clues to the sequence of events at Pontnewydd. The human occupation probably dates from an early part of MIS 7, the cold interruption may have led to the first mudflow event, and the stalagmites (which require warm and moist conditions to grow) probably formed with the return of interglacial conditions.

The sequence of fossil mammal bones in the mudflows suggests that human occupation occurred during relatively open conditions. Beaver, wood mouse and roe deer are represented, perhaps suggesting open woodland, followed by species such as lemmings, suggesting cooler, more open steppes. Much later still, long after the phase of human occupation, reindeer, musk ox and wolves roamed a treeless tundra. The mudflows contain hundreds of artefacts, and some are fresh enough to suggest that they have not been transported far. They include more than fifty relatively small pointed handaxes, Prepared Cores and flakes struck from them, and smaller tools that could have been used for working skins or wood, or as knives or spear points. The ancient inhabitants of Pontnewydd faced serious problems in their toolmaking, as there was no primary source of material like flint within 50 miles of the cave. They therefore had to rely on the rocks and pebbles that they could find locally, transported by rivers or by earlier ice flows, and the dominant stones were hard and volcanic in origin. This may explain why some of the tools look quite crude, and why the toolmakers seem to have made some knapping mistakes. Modern replication experiments show that such rocks are certainly more difficult to shape than flint, although reasonable results can be achieved with care and attention. Some of the tools show signs of having been heated, presumably as they lay in or near an ancient hearth, and this has allowed the heating event to be dated by the luminescence method to about 225,000 years ago, consistent with the age obtained from the overlying stalagmite. This dating method

OVERLEAF: A dramatic view of reindeer migrating through the snow at Nunavut, Canada. Scenes like this would have been common during the later British cold stages.

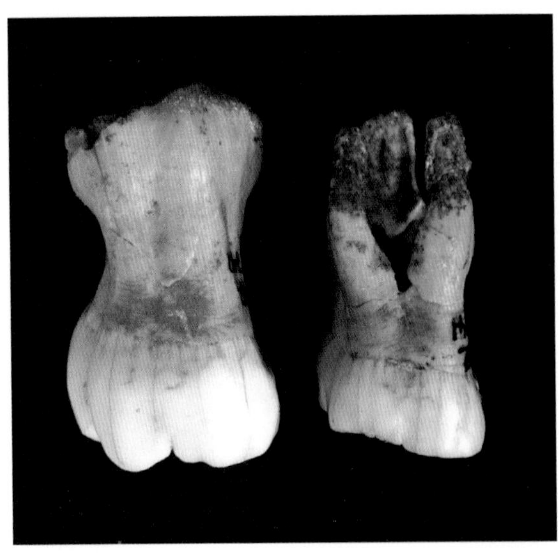

ABOVE: Two molars from a child's upper jaw found at Pontnewydd. One (*right*) is a milk tooth, while the undivided ('taurodont') roots of the first permanent molar are like those of some Neanderthals.

depends on measuring the amount of radiation damage that accumulates in an object when it is buried. The high temperature of a fire purges any previous radiation signal, allowing the time since the burning event to be estimated.

Amongst the bones and stones in the mudflows at Pontnewydd were seventeen human teeth, which I have been studying with Tim Compton, a colleague at the Natural History Museum. The teeth are from at least five people, and it seems likely that the tooth lost after 1874 and mentioned above was part of this series. They were probably more complete fossils before they were churned up, but only two of the teeth are now in their original sockets. Judging by tooth wear, three of the individuals were children aged around 12, one was a young adult, and the last was a mature adult. Many of the molars had lost their roots, but where they are preserved, several show a condition called taurodontism, where the roots merge for much of their length. This unusual shape is found in some modern humans, but in fossils it is really known only in Neanderthals, where about 50 per cent of their molar teeth show it. Its origin seems to be genetic and it may confer some advantages where teeth suffer heavy wear. A normal molar will cease to function as a chewing surface when the crown has worn down to root level and it disintegrates. However, in a taurodont tooth, the roots can fill with secondary dentine and remain functional for longer, as the disintegration is delayed. There are also features on the Pontnewydd tooth crowns, such as crests and cusps, that are reminiscent of Neanderthals, and their size and shape is similar to the samples from the Pit of Bones at Atapuerca, dated to about 400,000 years, and to Neanderthal teeth from the Croatian site of Krapina, dated to about 130,000 years. Overall they fit a European pattern of gradual 'neanderthalization' between about 400,000 and 150,000 years ago.

Given the redeposited nature of the Pontnewydd finds, it is difficult to reconstruct much more about the human occupation. Certainly the location was a good one, close to a river and raw materials in the form of cobbles and pebbles, and a wonderful vantage point from which to look for game in the valley. The cave entrance 250,000 years ago was probably not large, so it is likely that it provided shelter for only a small group of hunter-gatherers, although we do not know how many occupations are represented in the tools and human fossils. From the tools, it was more than just a sleeping place, with evidence for manufacture, resharpening, butchery and perhaps the working of skins and wood, as well as (by the implication of burnt tools) fireplaces. We have no clues about how these early Neanderthals died and became fossilized, or whether their bodies were left by the occupants or worked over by scavengers. Sites like Krapina show cut marks on the human bones suggestive (to some) of ancient cannibalism, but no evidence of this survives on the Pontnewydd fossils. But Pontnewydd is the most north-westerly handaxe site in the Lower Palaeolithic world, and its very remoteness and uniqueness tell us something very important. The lucky escape of the Pontnewydd evidence from the glaciers shows us what we must be missing on the uplands and valleys of western Britain due to the destructive effects of the ice ages.

In contrast, the southern lowlands of the Lower Thames Valley escaped glaciation completely, and two sites in Essex, West Thurrock and Aveley, have yielded important material from the time of Marine Isotope Stage 7, between about 200,000 and 250,000 years ago. An old tramway cutting at West Thurrock has been producing Middle Palaeolithic tools of Prepared Core type for over a century, as well as many shells and bones, while a quarry at Aveley became famous in the 1960s for its elephant and mammoth skeletons, now on show at the Natural History Museum in London. The sequence of river deposits of sands, gravels and muds is similar to the one already described at Purfleet, laid down by the Thames in the previous interglacial, 100,000 years earlier. However, the evidence from Aveley shows clear signs of a disturbance in the middle of the sequence that probably reflects the severe cold stage that interrupted the later interglacial. The earliest deposits at both West Thurrock and Aveley were laid down at the end of glacial stage MIS 8, about 250,000 years ago, but warm conditions soon followed, with woodlands and large mammals such as brown bear, rhinos and aurochs (extinct wild ox). Smaller mammals such as wood mouse, pygmy shrew, bank vole, common vole, water vole and beaver suggest rich vegetation, with rivers and ponds, as do the remains of shellfish and fish. The latter include tench, which require mean summer water temperatures not

lower than 20°C to spawn, and this is backed up by the presence of the European pond terrapin, which also needs warm waters to breed. The fish and shellfish together suggest stagnant or slowly flowing water with plenty of vegetation cover, typical of ponds, river backwaters or the lower reaches of a large river.

There is one particularly unusual feature of the large mammals at West Thurrock and Aveley. In Britain, the alternating conditions of glacial and interglacial are normally marked by alternations of certain species, such as reindeer and mammoth in the cold stages, red deer and straight-tusked elephant in the warm ones. However, in this interglacial the straight-tusked form is there as would be expected, but it is joined later by a species of mammoth as well. Moreover, there are many more mammoth fossils than straight-tusked elephant ones in the later deposits at Aveley, and at West Thurrock, so what does this mean? Well, the mammoth in question has small and distinctive molar teeth compared with the normal mammoths we find in the ice ages, and it seems to represent a less specialized form that grazed warmer grasslands rather than cold steppes. After the cold interruption, about 230,000 years ago, the forested vegetation of the Aveley interglacial never fully recovered and the dominance of grasslands seems to be reflected not only in the prevalence of these mammoths but in greater numbers of grazing rhinos and, now, horses. This suggests that new species such as the mammoth and horse migrated into Britain after the cold interruption, implying a physical connection with the rest of Europe at this time. These new species included a unique and exotic record for Britain: the jungle cat, which despite its name is most commonly found in grasslands and marshy ground near rivers in southern Asia today.

The Aveley interglacial is also represented in deposits further up the River Thames at Ilford in Essex, and in other areas such as Suffolk (Brundon and Stoke Tunnel), Buckinghamshire (Marsworth) and Oxfordshire (Stanton Harcourt). At this last site, in quarry deposits that may have been laid down in the upper reaches of the ancient Thames, about a thousand large bones, teeth and tusks have been recovered including those of bison, bear, lion, horse, hyaena, and once again, elephant and mammoth. Fifty species of shellfish and snails, ninety species of insect, and well-preserved remains of pollen, wood, nuts and seeds from oak, hornbeam, alder, hazel and willow paint a vivid picture of the environment there about 200,000 years ago. Grasslands, grazed by bison,

OPPOSITE: This mammoth jaw is one of many dredged from the North Sea in the collections of the Natural History Museum. It has mussel shells adhering to it.

horse and mammoth, bordered the river but there were also forested areas nearby, frequented by deer and elephants. A few stone tools give witness that humans were also there, at least occasionally. The Aveley interglacial inevitably drew to a close, and a site south of the River Thames at Crayford in Kent shows the transition into the long and severe cold stage of MIS 6. Crayford signals the oncoming glacial with the appearance of large mammals such as woolly rhino and musk ox, and small ones such as lemmings and ground squirrels. Humans hung on at Crayford for a while, for there are beautiful flint flakes made by the Prepared Core technique alongside woolly rhino jaws, but Britain was about to enter the longest phase of human absence seen for half a million years – an abandonment lasting over 100,000 years. During this time the nearest evidence of human occupation has to be sought across the Channel, in Jersey.

BELOW: The site of Crayford has Levallois flakes in direct association with woolly rhino jaws, suggesting that early Neanderthals may have lingered for a time in Britain as it got colder about 200,000 years ago.

Today the Channel Island resort of Ouaisné has golden sands lapped (on a good day) by a beautiful blue sea. But 150,000 years ago, its sea cliffs towered above a vast cold prairie where woolly mammoth and woolly rhinoceros roamed and early Neanderthals camped and butchered the carcasses of their prey. The nearly treeless prairie stretched towards England and all the way to France. Even a fall of sea level of 20 metres was sufficient to rejoin the Channel Islands to France, and at times of maximum glaciation the fall exceeded 100 metres.

ABOVE: Many holidaymakers at Ouaisné in the Channel Islands are unaware of the nearby site of ancient human occupation at La Cotte de St Brelade.

Excavations at a deep rift in the cliffs called La Cotte de St Brelade began in 1910, and culminated in major excavations during the 1970s, in which Prince Charles took part when a student at Cambridge. The digs have produced extensive evidence of early human occupation spanning a quarter of a million years, some 140,000 stone tools, and intriguing piles of mammoth and rhino bones dating from about 150,000 years ago. On at least two occasions, separated by thousands of years, it seems that the early Neanderthals at La Cotte drove herds of mammoth, accompanied by a few woolly rhinos, over the cliffs or into the chasm to their deaths. We do not know how they did this, but extensive evidence of burning in the site raises the possibility that fire was involved, either the firing of dry vegetation or the use of torches. Nevertheless, stampeding the herds in this fashion would have required planning and teamwork, as well as bravery.

The Neanderthals must have dismembered the carcasses into chunks that they could carry or drag into the shelter of La Cotte, and bones, skulls and tusks were probably piled up against the walls of the chasm as they finished working on them, or to keep access and floor-space free. Whereas they normally returned to the site repeatedly, on these two occasions people did not come back for some time, and dust storms blanketed the bone heaps, protecting them for the next 150,000 years.

The stone tools found at La Cotte vary from small flake tools early on, through assemblages dominated by scraper tools (for working skins?), to the kinds of tools made by late Neanderthals in neighbouring Brittany. Microscopic study of the flake and scraper surfaces shows that some have what looks like hide polish, while notched tools occur in different frequencies, which may reflect the differing availability of trees in the surrounding landscape. The frequency of possible spear points also generally increases through time, with particularly sturdy examples being manufactured in some of the levels with bones of megafauna such as mammoth and rhino, perhaps an indication of a requirement for strength. Whereas the first inhabitants of La Cotte seem to have had abundant supplies of flint, tools in the later levels show much greater

BELOW: (*left*) The huge cleft in the cliffs at la Cotte de St Brelade, where early Neanderthals sheltered 150,000 years ago. (*right*) The excavation team who discovered Neanderthal teeth at La Cotte in August 1910.

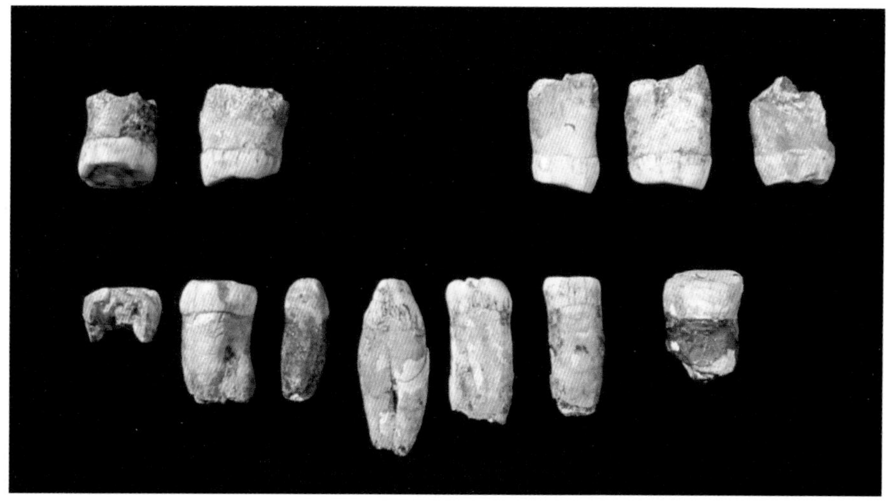

LEFT: The Neanderthal teeth from La Cotte, in which the unusual root shape known as taurodontism was first recognized. These are the only late Neanderthal fossils so far found in the British Isles.

evidence of reworking and rejuvenation, suggesting that raw material was becoming scarcer. An interesting by-product of archaeological study of the direction of the resharpening processes is a demonstration that it was predominantly carried out by right-handers in a ratio of 4:1, close to the normal ratio of right- to left-handers in people today.

As already mentioned, there is extensive evidence of burning throughout the deposits of La Cotte, in the form of burnt tools and stones, charcoal, bone fragments and comminuted bone powder. Its presence seems to correlate with the presence of stone tools and butchered bones, suggesting that it is directly associated with human activity, rather than the result of natural fires, and as such is one of the earliest good examples of this behaviour in Europe. However, no hearths were preserved in the lower levels, either because fireplaces were unstructured, or because they have been reworked by later activity. But what is especially unusual about the La Cotte evidence is that burnt bone fragments greatly outnumber those of wood, suggesting that in relatively treeless landscapes the Neanderthals resorted to burning bones. This may seem an unlikely practice to us, but people who live in similar environments today, such as the Inuit, regularly burn bones for fuel. Mammoth and rhino bones would have had a high fat content, and would have burnt well once they were at a high temperature, implying that kindling would have been needed to start the process.

There was a high ridge or plateau of chalk rock connecting Kent with the French region of Artois

OPPOSITE: The Straits of Dover and the English Channel, as seen from space.

So if people (presumably early Neanderthals) were in Jersey 150,000 years ago when the sea level was low enough to connect what are now the Channel Islands to France in one direction, and to southern Britain in the other, why didn't they regularly cross what is now the bed of the Channel into England, if it was dry land? This brings us to one of the key questions in understanding the ancient human occupation of Britain: when did Britain become an island for the first time? This question is not as straightforward as it might seem, because the creation of Island Britain was not just a matter of the sea level becoming high enough to cut Britain off from continental Europe. As we explained in Chapter 1, there was a high ridge or plateau of chalk rock connecting Kent with the French region of Artois, and until this was cut through, sea level rises on their own were not enough to isolate Britain. So what happened to breach the chalk, and when did it happen?

If we go back to the Pliocene period, over 2.5 million years ago, Britain seems to have been an island then, surrounded by mild seas. Although the Antarctic continent had been ice- and snow-bound for millions of years, there was little ice at the North Pole and in Greenland, and so sea levels globally were much higher than today. But around 2.5 million years ago, changes in the Earth's oceanic and atmospheric circulation patterns started a permanent build-up of ice and snow in the north, which led to a sea level fall, and the gradual exposure of a plateau of land around the chalk ridge between Britain and continental Europe. In addition, European rivers flowing into the region of the present North Sea extended their headwaters and dropped increasing amounts of sediment into the southern North Sea basin, gradually increasing the area of land connection so that Britain was a peninsula of Europe by about 1.8 million years ago. This was still the situation when massive ice sheets started to spread across northern Britain in the glacial stages after 700,000 years ago. Geological evidence shows that a huge freshwater lake built up in what is now the southern North Sea towards the end of the severe Anglian glacial stage, corresponding to Marine Isotope Stage 12 (about 450,000 years ago). The chalk ridge formed the southern

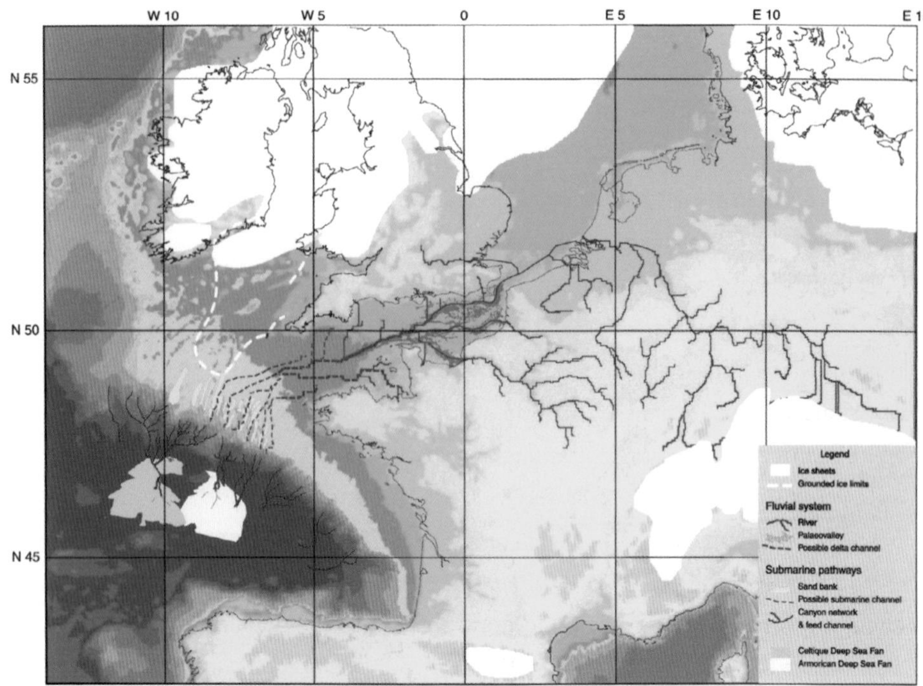

RIGHT: Surveys of the Channel using radar, sonar and geological coring are building up an increasingly detailed picture of a huge and now-submerged system of river valleys. This reconstruction for the peak of the last glaciation was produced by French researchers in 2003.

shore, but eventually the lake overflowed and cut through the chalk to form the Strait of Dover. Other workers, however, have suggested that the event was much later, and that the Strait was still closed during MIS 7, 200,000 years ago. In which case the Channel could have been cut during the succeeding cold stage.

A team of researchers at Imperial College, led by Sanjeev Gupta, has been using the results of advanced sonar investigations to survey the sea floor several miles off the Sussex coast, in order to study how the Channel formed. A special echo sounder, strapped to the hull of a survey boat, has revealed a landscape hidden beneath the waters for ten thousand years. During the low sea levels of the glacial stages the River Arun, which now enters the Channel at Littlehampton, ran on for several miles before flowing into a valley carved by a massive south-flowing river created by the combined waters of the Thames, Rhine and Seine. In places this huge underwater valley is more than 10 kilometres (7 miles) wide and 50 metres deep, with vertical sides, and its nearest geological parallels are found not on Earth, but in the monumental flood terrains of the planet Mars. This suggests that the valley was not formed by normal geological processes, but was indeed created by catastrophic flood flows following the breaching of the Dover Strait, and the sudden release of water from a huge ice-dammed lake to the north. Gupta believes that the creation of the English

Channel was one of the most powerful flood events known on Earth. However, the timing of the event is still not clear, and it could even have happened in two stages, perhaps even in two separate glacials. Clues to the timing of the breach could come from studying geological cores produced when the Channel Tunnel was being built, and from sampling deposits in and around the huge submerged river system, and this is something Gupta and AHOB are pursuing. But knowledge of the scale of the Channel river also changes our perceptions of Britain's isolation from the Continent. The old view that the Channel formed a barrier to movement only in interglacials must be modified – even at times of lowered sea level this deep and wide river could still have been a formidable obstacle to the movement of animals and, without boats, humans.

This irreversible alteration in geography might have combined with changing environments to make Britain gradually less suited to human occupation. It probably took a long time for humans to adapt properly to European environments, and to make things worse the environments themselves gradually transformed through time as climates became more extreme. The Neanderthals in north-west Europe increasingly adapted to cooler and more open environments and this is reflected in a reversal of the earlier situation, where it was the interglacials that showed the densest occupation. For example, the remains of mammoth, reindeer and horse dominate German Middle Palaeolithic sites, and in the briefer warm stages the Neanderthals seem to have preferred to migrate eastwards to the more open steppes. The distribution and movement of herds in the steppes would have required greater human mobility and more portable technology, which is perhaps why the Prepared Core technique became more common, and people left less stone tool debris behind. If Britain had less of the preferred steppe environments, particularly in later interglacials, then the Neanderthals may have had a hard time surviving there, particularly if their mobility was restricted by the growth of a massive Channel river system in the colder periods, and the Channel itself in the interglacials.

There are clues in biology and archaeology indicating when the breach between Britain and continental Europe occurred, and to its scale. By looking at the warmest stages of the interglacials, when ice caps should have been at their smallest and sea levels at their highest, we can see that there are greater

ABOVE: William Buckland, again dressed for fieldwork.

OPPOSITE TOP: Victoria Cave is located in the Yorkshire Dales, where hippos roamed 120,000 years ago.

similarities between these neighbouring regions in both mammals and molluscs (land and water-living) during the warm stage 400,000 years ago than the warm stage 125,000 years ago, suggesting the breach was either absent, or less significant, during the earlier warm stage. The later warm stage, the last interglacial (if we don't count the one we are living in today), is known in Britain as the Ipswichian, after the site in Ipswich where its distinctive pollen signal was recognized. Most experts accept that Britain was an island by then because it was home to an idiosyncratic combination of mammals known as the Joint Mitnor Cave Mammal Assemblage-Zone, after a site in Devon. This unique assemblage is found nowhere else at the time. It was first excavated from Kirkdale Cave in Yorkshire in 1821 by William Buckland, and this is one of several sites in the region where hyaenas seem to have been scavenging hippo carcasses at lakes or rivers. At Victoria Cave, 400 metres up on the now craggy Yorkshire Dales, we have one of our best fixes on the age of this odd fauna, since fossil bones are enclosed in stalagmite that has been dated to about 120,000 years ago by the uranium-series method, which relies on the steady decay of radioactive uranium into its daughter products. And Buckland was right in one of his major observations about this 'hippo fauna' – there are never any associated human relics (another contrast with the preceding four interglacials, which always have some evidence of humans).

At first glance much of mainland Britain, from Yorkshire southwards, would have looked to us like a wooded African game park with lion, hyaena, hippo, elephant and rhino. But on closer inspection, more familiar species like red deer, fallow deer, water vole, wood mouse and field vole would also have been seen, although horses were absent, unusually for an interglacial. Most of the species present must have been fast movers, for the warm start of this interglacial seems to have been rapid, along with sea level rise, as the previous glacial ice melted. Plants, insects and birds would have been amongst the first colonizers from Europe, followed by herbivorous and then carnivorous mammals. Some large mammal species could even have swum across. We know that hippos were happily wallowing in the Mediterranean at this time, so once the ocean was warm enough they could simply have swum along the Atlantic coasts towards Britain, and then swum up the rivers (including the Thames, where their bones have been excavated from the foundations of buildings on Trafalgar Square). If we assume that hippos in Britain behaved similarly to the living species in Africa, herds of up to thirty of these enormous animals would have wallowed together in shallow rivers or lakes. In the evening they would have emerged to feed separately, and they do feed on an enormous scale, sometimes travelling several miles in a night. A hippo can graze for about six hours and consume up to 100 pounds (45 kilograms) of grass in that time, and so even a single herd could have made quite an impact on the vegetation around British rivers and lakes.

Plants and insects suggest that the climate was at least one degree warmer than today, with mixed oak and hazel forests and some southern exotics such as the Montpellier maple and water chestnut. Bird faunas show the same pattern, with Mediterranean species such as Cory's shearwater nesting on British coastlines. At this time the subtropical Gulf Stream was flowing northwards

BELOW: The tusk (canine) of a hippopotamus that swam in the Thames, excavated from the foundations of Uganda House, Trafalgar Square.

ABOVE: A vivid reconstruction of Trafalgar Square as it may have looked 120,000 years ago.

OPPOSITE: The strange assemblage from Banwell Bone Cave includes a giant bear, probably an even more formidable predator than the polar bear. The top of its humerus (*right*) is compared with the whole humerus of a brown bear (*left*).

particularly strongly for about 15,000 years, and the existence of the Channel would have facilitated the spread of these mild Atlantic waters through the North Sea, towards the Baltic. This may well explain why it was not just Britain that was balmy at this time. Fossil pollen suggests that hazel and alder were growing as far north as Swedish Lapland, and from the evidence of ancient shorelines Scandinavia was probably an island too. Ireland must have been similarly sultry, but surprisingly little evidence has so far emerged from there.

About 115,000 years ago, that balmy flow across the Atlantic suddenly switched off as the oceanic conveyor belt that transfers subtropical warmth northwards reversed, and Britain's climate started an erratic downturn into the last glaciation. Sites that we have excavated in the Gower Peninsula such as Bacon Hole and Minchin Hole caves show this decline about 90,000 years ago, with the disappearance of some warmth-loving species and the appearance of cooler indicators such as mammoth, and northerly birds such as the long-tailed duck and bean goose. Some interglacial species, elephant and narrow-nosed rhino, still hung on in this assemblage, known as the Bacon Hole Mammal Assemblage-Zone. But 15,000 years later they had finally all gone, as evidence from Banwell Bone Cave shows. We encountered this Somerset site in the Introduction, where it figured as a solemn reminder of the Flood, a sort of biblical deluge theme park. And it is easy to see why people in 1825 were so impressed by what they saw. William Beard, a local collector, had found so many fossil bones in the cave that he piled thousands of them in decorative stacks against the walls. The great majority were of bison and reindeer, with rarer species such as wolf and brown bear. The bear remains are often huge, suggesting a powerfully built runner with small teeth, and enormous claws, an active hunter like a polar bear. The wolf skulls and jaws have heavily worn and broken teeth, suggesting that this species might have been filling the role of scavenging and bone crunching normally taken by the hyaena, of which there is no trace at this time. The assemblage of mammals is so distinctive that it

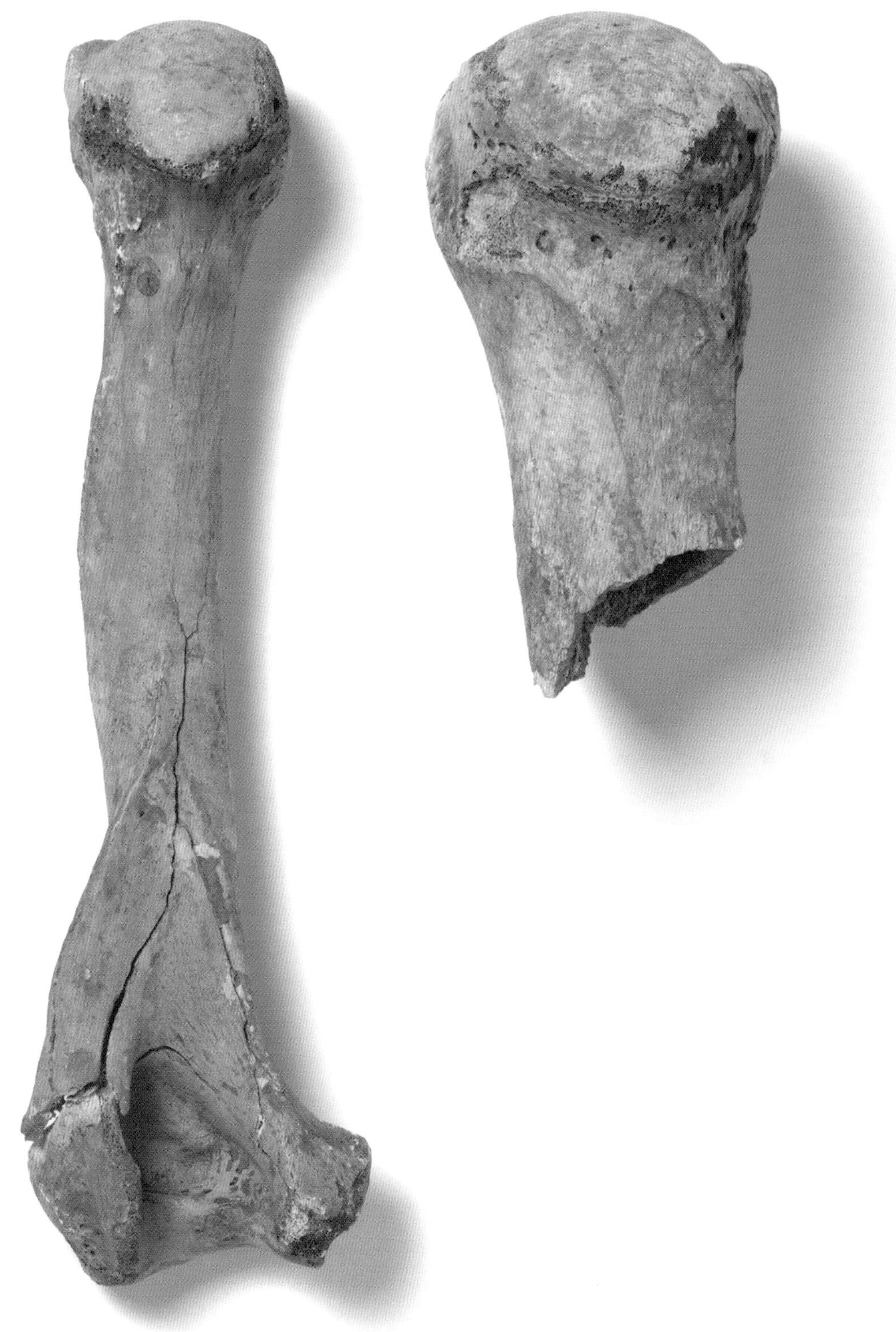

ABOVE: William Beard (1772–1868), who managed the Banwell Estate, collected bones from a number of local sites. He was still taking visitors to the Bone Cave at the age of 93.

can be recognized at a number of other British sites, all of them placed in the Banwell Bone Cave Mammal Assemblage-Zone, and representing a cold stage about 75,000 years ago, with forms like arctic hare, arctic fox and wolverine. As well as hyaena, there is once again another notable absentee from the Banwell mammals: Man. There is not a single fossil human or archaeological relic from this time, continuing the pattern of the previous interglacial.

The limestone in which the Banwell Bone Cave formed is an outlier of the Mendip Hills, and there are many similar bone collections dominated by bison and reindeer in the region from caves, gullies and open slope deposits. So what does this signify, and how did they accumulate? Bison and reindeer both form large herds, so could the animals have been trying to shelter in the cave, or could they have fallen through a hole in the roof? This seems unlikely, firstly because the cave has a relatively narrow entrance and small chambers, and an unbroken roof. But more importantly, even from the initial crude excavations, it was seen that the bones at Banwell were almost always jumbled up; neighbouring ones could not readily be reassembled into skeletons, so the carcasses were already decayed and broken up by the time they were deposited.

The most likely scenario, based on similar events today, is that massive herds of bison and reindeer were migrating across a landscape with deep snow, which obscured dangerous ground – a cliff, a ravine, even a particularly deep pocket of soft snow. Many of the animals became trapped, died, and decayed. In the spring melts, the bones fell or sludged into the cave, and as long as the landscape remained unchanged, the process could have been repeated over many winters.

The environmental picture we get from Banwell and sites like it is an unremittingly bleak one, a cold and windblown Britain similar to northern Scandinavia today, untouched by human presence. Yet other sites from around this time such as Cassington, near Oxford, and Isleworth in West London have beetle remains suggesting that it was sometimes much warmer, and yet there were still no signs of people. The Neanderthals were certainly across the Channel River in Belgium and France, and were occupying La Cotte de St Brelade in Jersey, where they lit fires and even left their remains in the form of thirteen teeth – the only late Neanderthal fossils from the whole of the British Isles. Did they cross the Channel River, and we have not yet found or recognized the evidence, or were they unable to? These are puzzles that AHOB is still investigating. Whatever the answers, within 15,000 years of the severe conditions represented at Banwell, people were definitely back in Britain, perhaps the first human inhabitants for over a hundred millennia. They were Neanderthals, and they occupied midland and southern Britain for much of the next 30,000 years, although there were many changes and challenges to come, and not just from the climate.

LEFT: One of the remarkable stacks of bison and reindeer bones in Banwell Bone Cave.

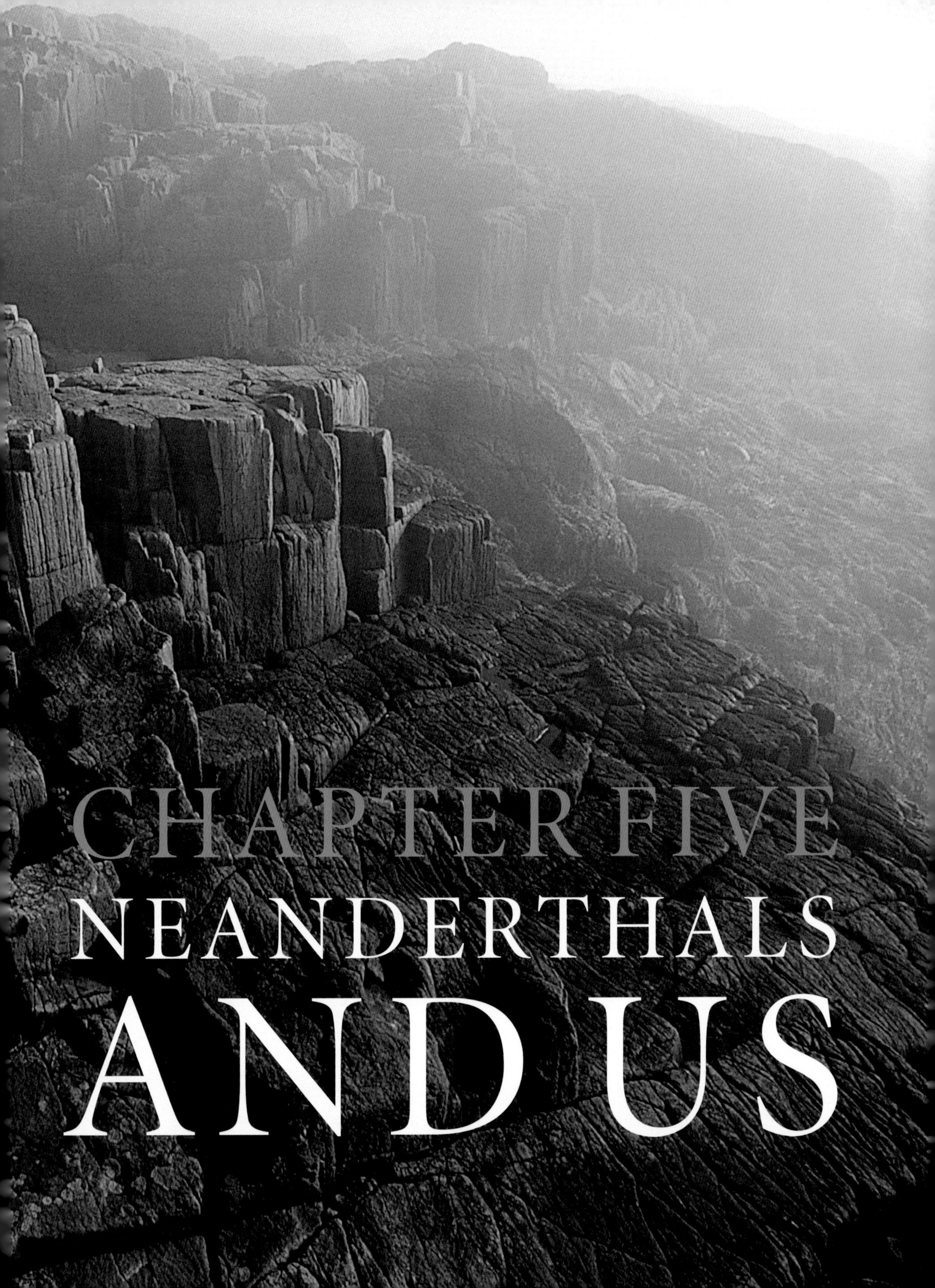

CHAPTER FIVE

NEANDERTHALS AND US

PRECEDING PAGES: Rocky outcrops like this would have been used by Neanderthals where there were caves for overnight shelter. Many caves in the uplands of southern Europe have evidence of Neanderthal habitation, and would have provided excellent vantage points to track the movement of game.

The Neanderthals have had an image problem over the years, and brutish behaviour is often still called 'Neanderthal'. Before Africa yielded up really ancient potential ancestors, the Neanderthals were often pushed into the position of 'missing links', portrayed as hairy and bestial, with stooped gait, long arms and grasping big toes. Yet in the 1970s the pendulum had swung so far in the opposite direction that they were instead depicted as 'flower people', and our immediate ancestors. The American anthropologist Carlton Coon famously remarked that if a Neanderthal could be washed and shaved and dressed in a business suit, he would pass unnoticed on the New York subway (I usually comment that this probably says more about the New York subway than it says about Neanderthals!). But research over the last ten years is at last giving us a realistic view of these fascinating predecessors as close relatives who were as human as we are, but in their own unique way. Over a period of several hundred thousand years in Europe and western Asia, the Neanderthals developed their own bodily (and no doubt behavioural and social) characteristics, in an evolutionary history largely separate from our own.

Because, like us, they had developed the habit of burying their dead, their remains in caves have been saved from erosion and damage, and lay awaiting excavation by archaeologists. Unfortunately only scraps have been found so far in Britain, but using more complete fossils from sites ranging from Gibraltar to Uzbekistan we can build a detailed picture of their bodies and constitution, how they grew, lived and died. They were relatively short, wide-shouldered, wide-hipped and barrel-chested – yet their posture and gait was essentially the same as ours. Their build looks more suited to short powerful bursts rather than endurance running, which perhaps fits with the idea that they were mainly confrontational hunters, getting to grips with their prey at close quarters, using short-throw or thrusting stone-tipped spears. In turn, this may explain why their skeletons not only show plenty of wear and tear from a demanding life-style but also healed wounds and fractures, particularly in the upper part of the body and head. This injury pattern is matched quite closely by that of rodeo riders, who also regularly have to confront untamed large mammals (although personally I believe some of the injuries could also reflect a high level of interpersonal violence).

If anything, the Neanderthal skull was even more distinctive than the body that went with it. The braincase was large, but long and low, without the domed forehead of modern humans. Within it was a brain of very large size, but of unknown quality compared with our own. Neanderthal faces were dominated by an enormous nose (high, wide and projecting), accentuated by sweptback cheekbones and a receding chin. The large eye-sockets were overshadowed by a double-arched and prominent brow ridge. This visage is so idiosyncratic that it can be recognized even in fragmentary fossils, and the beginnings of its development can be seen in the youngest Neanderthal children.

The Neanderthals are found across Europe and the Middle East, and they evolved through a time when the climate was dominated by cold. However, although their short and stocky body shape seems suited to the cold, they were not just people of the ice ages. They are certainly found fossilized with reindeer and mammoths 50,000 years ago in Germany, but also alongside elephants and hippos 120,000 years ago in Italy. Yet, as we have seen, they did not get to Britain when that kind of fauna spread here during the last (Ipswichian) interglacial; they must have lacked the capability to cross the nascent English Channel.

ABOVE: The skulls of a Neanderthal (La Ferrassie 1, *left*) and Cro-Magnon (Cro-Magnon 1, *right*).

Instead we have to wait until about 60,000 years ago for the first traces of their return in over 100,000 years, and that evidence takes us back to Norfolk. And what a different scene it now was from the balmy climate of Pakefield at the start of our story, some 650,000 years earlier.

Amateur archaeologist John Lord lives in Norfolk and, with his wife Val, demonstrates Stone Age life, wearing skin clothing and making and using replicas of ancient artefacts. For several years he kept a watching brief on a working gravel pit near the village of Lynford, since it had occasionally produced fossil bones and Palaeolithic tools. But in 2002, the pace of discovery suddenly accelerated, and it was clear that the quarrying had reached a rich seam of mammoth bones. Soon afterwards he and AHOB Associate Nigel Larkin found the first of numerous beautiful small flint handaxes. English Heritage agreed to fund the Norfolk Archaeological Unit in mounting a rescue dig before the evidence was lost to science. AHOB members and associates took part in the excavations and have been involved in the work at every stage, and Lynford has developed into one of the most important Palaeolithic sites in Britain. Approximately 2000 kg of samples were removed for processing and a rich harvest of pollen, plants, molluscs, insects and vertebrates has been recovered.

Sixty thousand years ago, an ancient forerunner of the modern River Wissey snaked over the landscape, laying down gravels and silts, in turn capturing and abandoning ponds and lakes. A channel fluctuated between flowing and still water, and on its banks and in its waters the bones of twelve species of mammal, fish, and several birds and amphibians accumulated, along with an astonishing assemblage of 160 species of beetle. Of these, twenty-one are not found in Britain today and can be found as far away as Siberia, but they included dung beetles and some carrion beetles, suggesting that large mammals were using the channel/lake as a watering hole, and that some were also dying there. The mammal remains were dominated (over 90 per cent) by the bones, teeth and tusks of nine juvenile and adult mammoths, but also represented were woolly rhino, reindeer, horse, bison, fox, wolf, hyaena and brown bear. The plant fossils showed that the landscape was mostly grassland and acid heath or bog, with at best a few birch trees, while the lake was bordered by sedges and bulrushes. However, plants like bilberry, chickweed, meadow rue, thistle and dandelion paint a somewhat less bleak picture. Combining the environmental evidence from beetles and snails suggests that the average summer temperature climbed to 13°C, several degrees colder than Norfolk today, while winters were much more severe, averaging well below –10°C.

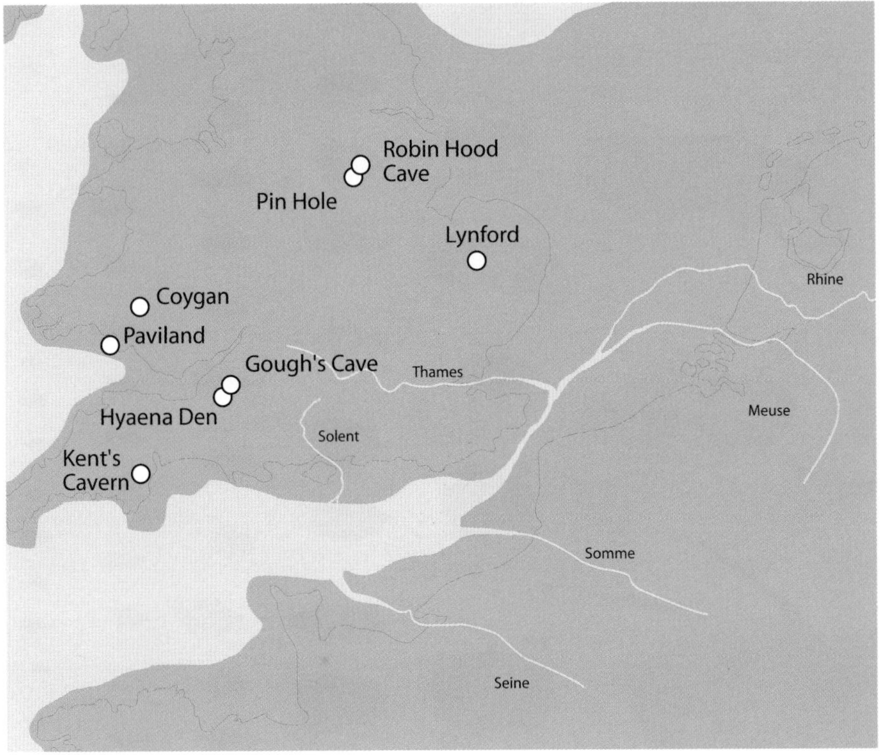

LEFT: A reconstruction of Britain as it may have looked about 50,000 years ago (sites covering a wider time range are shown). The English Channel had formed by now, but at a time of lowered sea level it was occupied by the Channel River system.

Remarkable as the environmental evidence from Lynford is, the significance of the site rests on the combination of mammoth bones and stone tools, the only such association yet found in Britain, and rare across the whole Neanderthal world. There are nearly 500 artefacts, including forty beautifully made small handaxes in shiny black or dark blue flint, and flakes that are mainly from the final stages of manufacture. This suggests that the Neanderthals were travelling to the area with nodules of flint that had already been knapped close to their final form, and they then produced exactly what they needed on the spot. The tools show quite a high incidence of damage and breakage, which perhaps reflects their use on recalcitrant large mammal bones. We can imagine small Neanderthal groups moving westwards across Doggerland (land now submerged by the sea), tracking reindeer herds as they migrated along rivers into Britain, and finding themselves at Lynford. At times wind chill would have been severe, and the lack of trees and therefore wood for fuel or shelters would have posed severe problems that they had to solve in order to survive. Perhaps they lit fires of dried reeds and mammoth dung, and like the occupants of La Cotte de St Brelade burnt mammoth bones when the fires were sufficiently hot. Like Neanderthals in Russia, they may even have used mammoth bones and tusks, covered by animal skins, to produce basic windbreaks. And it is difficult to think that they did not have at least rudimentary skin or fur clothing and snowshoes to cope with such bleak conditions.

So what were the Neanderthals actually doing at Lynford? That is a more difficult question as, although some of the handaxes were found amongst mammoth bones, none of the bones show direct evidence of butchery in the form of cut marks. In itself this is not surprising, since modern butchery experiments on dead elephants in zoos and the wild has shown that, with the sheer mass of meat, tools may never cut through to the bone. However, some of the Lynford bones show impacts and breakage that occurred when they were fresh, and some woolly rhino and reindeer teeth show fractures that suggest the jaws may have been split to extract marrow. In addition, compared with the numbers of mammoth skulls, teeth, jaws and tusks, there is a distinct lack of meat-bearing limb bones – had they been carried away by the Neanderthals for meat or fuel? Further study may throw more light on these questions, but we can guess that in difficult times the Neanderthals could not have afforded

OPPOSITE: A collection of modern beetles at the Natural History Museum. Beetle remains are common at Lynford, and provide vital clues about the climate and environment.

RIGHT: Lynford in Norfolk has an association of mammoths and Neanderthal archaeology that is unique in Britain. Here the huge tusk of an adult mammoth is under excavation.

to turn their noses up at large quantities of meat, in whatever form it came. Perhaps they were actively hunting mammoth, as is suggested by the discovery of a yew spear tip amongst elephant ribs at the last interglacial site of Lehringen, in Germany. But what is also interesting is that many of the Lynford mammoth bones show evidence of disease or injury, particularly in the ribs and backbone. Were these weaker animals that had died near the Lynford lake or, because of their weakness, were they sought out as prey by hyaenas, wolves or Neanderthals? If the Neanderthals were well organized they could certainly have got their share of the plentiful resources provided by a dead mammoth or rhino, and perhaps they even took advantage of a natural cold store, returning regularly to the lake to exploit frozen or thawing mammoth steak.

ABOVE: A bear molar from Lynford.

OVERLEAF: This stuffed mammoth from the Siberian permafrost was excavated at Yakutia in 1903, and exhibited in the St Petersburg Museum of Zoology. There is still an extensive trade in mammoth tusks from the region today.

The evidence from Lynford is in line with work on Neanderthal diet carried out by AHOB members, and obtained through analyses of carbon and nitrogen isotopes in fossil bones. The method is based on the principle of 'you are what you eat', since the foods we eat leave their characteristic chemical signatures in the bone we lay down during our lives. If bones are well enough preserved they retain these signatures, which can be read through laboratory analyses to indicate the proportion of plant, meat or fish foods consumed by the owner of the bones before they died. Neanderthals in central and northern Europe certainly appear to have been highly carnivorous and at the top of their food chains, comparable with beasts such as the hyaena, and they would have needed to be in places like Lynford, when plant resources would not have been sufficient for much of the year. But further south, on the Mediterranean coasts of Italy and Gibraltar, we get a very different picture of Neanderthals apparently happily subsisting on a diet of baked tortoise, rabbit, and seafood such as mussels. Clearly they were resourceful and adaptable people.

The deposits at Lynford have been dated by the luminescence method to about 60,000 years ago, and we have no sound evidence of the predecessors of these people in Britain after about 200,000 years ago. For that we have to cross the Channel River system to France, or Doggerland, to sites in northern Europe where we know the Neanderthals did survive for at least parts of the intervening period. But in the brief windows of glacial warming (if Lynford can be called that), the Neanderthals made their comeback, and soon after this we find them associated with similar animals and a comparable climate in the Creswell region

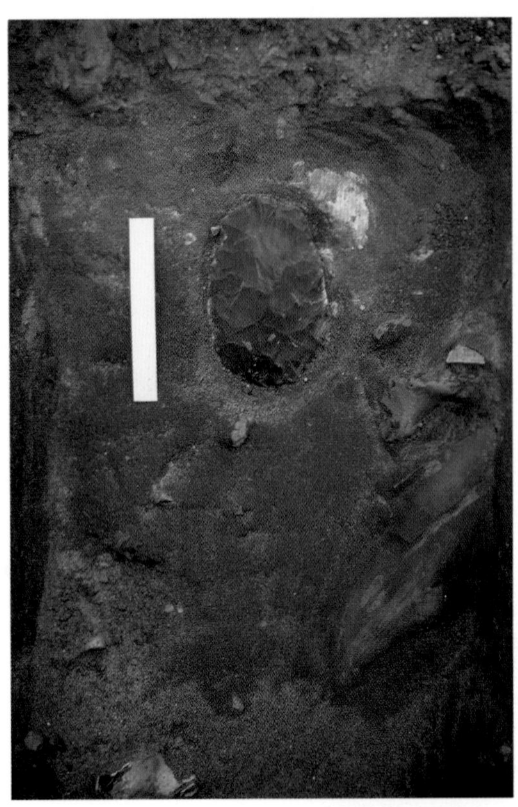

ABOVE: One of forty beautifully made small handaxes from Lynford.

of Derbyshire, at Pin Hole and Robin Hood Cave, in Hyaena Den Cave at Wookey in Somerset, and far across Britain in Coygan Cave, Carmarthenshire. We get further glimpses of their occupations at classic sites like Kent's Cavern and Paviland, usually in the form of small handaxes with rather flat butts. These are similar to French examples from what has been called the Mousterian of Acheulian Tradition (though a lineal connection to much more ancient handaxe industries is very doubtful), but this may indicate Neanderthal connections between Britain and south-west France at this time.

Things started to change about 40,000 years ago. New stone tools known as leaf points, because of their lanceolate shape, appeared in Britain, and instead of similarities with France, the most obvious parallels are found in Germany and Poland. And 35,000 years ago, there were further changes, on an accelerating scale. Even though the Neanderthals were surrounded by bone, antler and ivory, they made little use of these materials to make tools, since they are hard to work without pre-treatment or specialized tools like chisels, burins or gouges. Yet after this period, bone points start to appear in British sites, along with the small stone tools used to make them. These specialized manufacturing tools were made on long, thin blades of flint, and they mark the arrival of the Upper Palaeolithic in Britain, soon after it had spread across Europe. This watershed in human behaviour was also marked by the appearance of the first known representational art in the form of carvings, engravings, and painting on cave walls. Beautiful little figurines of animals and strange humans with lion heads were made in Germany from the intractable medium of mammoth ivory, along with flutes manufactured from swan bones, while deep in the cave of Chauvet in France, whole processions of woolly rhinos, lions and horses were brilliantly drawn in charcoal. And in Goat's Hole Cave in South Wales, a man was buried with red ochre pigment and ivory ornaments, whose partial skeleton would be excavated 27,000 years later by William Buckland and mistakenly called The Red

LEFT: The muds and silts at Lynford contain well-preserved remains of insects, plants and wood.

65
3-22
14.
E.4846
65
3-22
15.

Lady of Paviland. He was a modern human like us, and part of the population known as the Cro-Magnons, after the French discovery of 1868. They first appeared in Europe about 40,000 years ago, and physically they were essentially like us, but somewhat larger bodied and (like Neanderthals) somewhat larger brained too. Compared with Neanderthals, they were taller and more linear, with narrower hips and shoulders, and body proportions more like people who live today in hot dry conditions – perhaps a clue to their ultimate origins. The first of the Cro-Magnons must have encountered their long-lost cousins, the Neanderthals, as they spread across Europe, and what happened in those meetings has inspired novelists and fuelled scientific arguments for over a century.

The first fossil human ever excavated systematically, from Paviland, was also the earliest known example of a particular burial tradition practised by the Cro-Magnons who made the Gravettian culture across Europe between about 25,000 and 30,000 years ago. The body was placed extended and close to the cave

OPPOSITE: Neanderthals had varied tool kits including small handaxes and tools with a curved scraping edge, perhaps for working skins.

SECTION OF THE CAVE CALLED GOAT HOLE.
In the Sea Cliffs 15 Miles West of Swansea

LEFT: Buckland may have excavated the first fossil human skeleton, from Paviland, although he did not realize it at the time.

OVERLEAF: Beautifully observed lions, drawn at Chauvet Cave in France over 30,000 years ago.

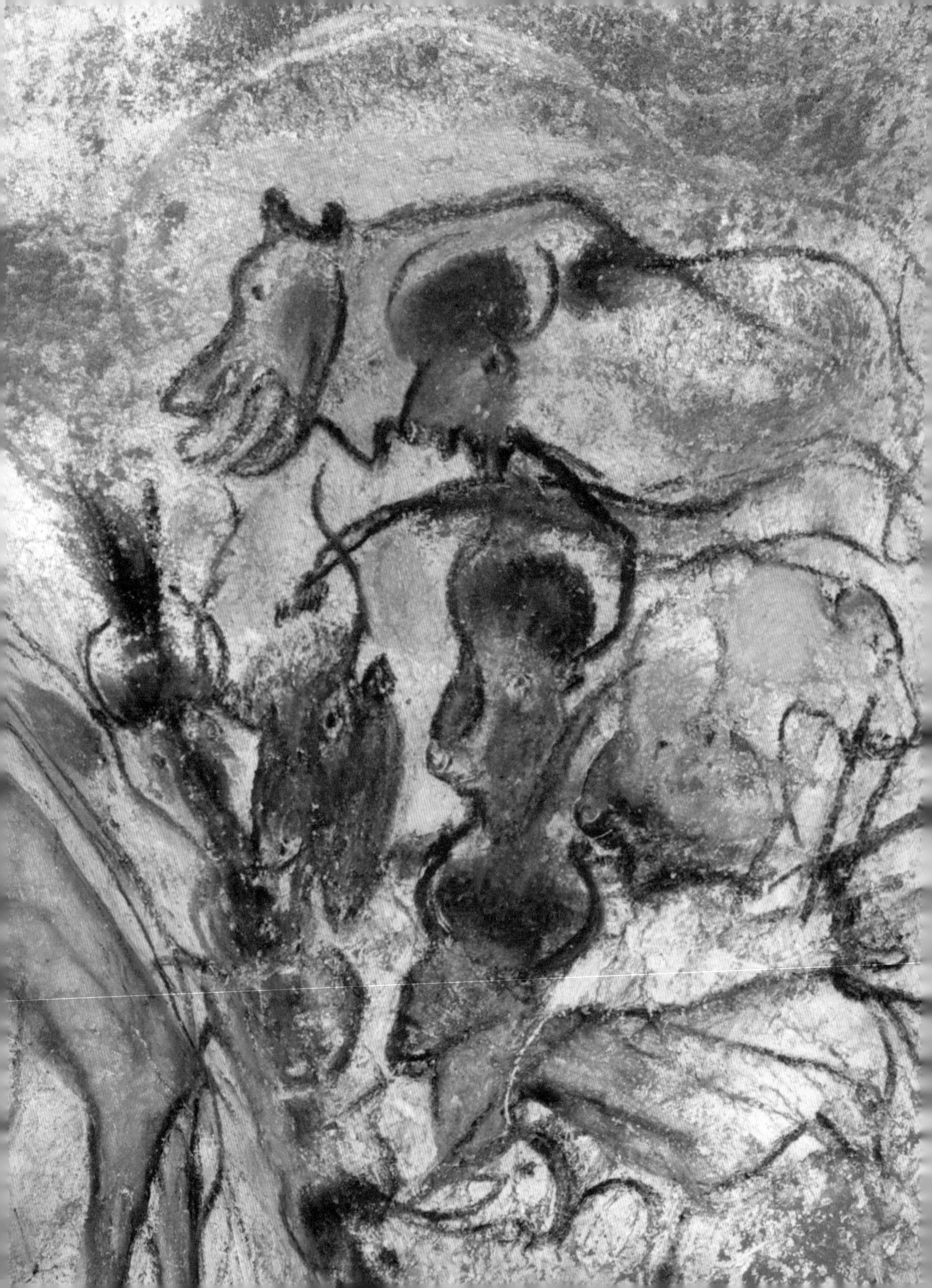

ABOVE: A figurine in ivory from the Ukranian site of Avdeevo.

wall, with rods and bracelets of ivory, and perforated periwinkle shells that were probably once part of a necklace. A mammoth skull and stone slabs may also have been placed there at the same time, although this is unclear now, and we do not know if the red ochre was painted on the skin or clothing of the man, or was sprinkled over the body as a final gesture. The man in question was a young and apparently healthy adult but at some way short of 6 feet (1.90 m) was relatively small by the standards of strapping early Cro-Magnon men. His skull was also missing, perhaps because of erosion by the encroaching sea. At about the same time as the Paviland man was interred, a thousand miles across Doggerland to the east the bodies of three people in their late teens were buried in mysterious circumstances at what is now the village of Dolní Věstonice in the Czech Republic. The oldest of the three, and the first laid in the grave, was in the centre, and suffered from severe growth abnormalities that deformed the backbone, hip bones and the legs, so much so that determining sex (which can usually be done reliably from the pelvis) is difficult. On the right was a large and robust male, lying face down, with the head turned away. His left arm covered the hand of the disabled individual in the middle, as if holding it. On the left was a smaller robust male lying on his left side and facing the individual in the middle, with both his arms reaching down to the pelvic region of the central individual.

The three skulls and the surrounding soil were impregnated with red ochre and there were rich head coverings of wolf and fox teeth and ivory beads. The central individual also had considerable quantities of ochre under the pelvis and between the legs. Further clues to the significance of the triple burial may come from a wooden stake apparently pushed into the body of the male on the left, and the fact that the male on the right might have had a smashed skull. Many different theories have been advanced to explain this strange burial pattern including murder, sacrifice, failed childbirth, and heterosexual or homosexual liaisons, and the burial has even featured in the novels of Jean Auel. A little later and even further east, two children and a man were buried at Sunghir in Russia, accompanied by thousands of mammoth ivory beads (probably sewn on clothing), hundreds of arctic fox teeth pendants, and a huge array of ivory ornaments, tools and weapons. These burials contained items that had taken days, weeks or even months of preparation, and were carrying important social messages that we cannot read now. They indicate that Cro-Magnon society was already complex and stratified in terms of roles and social status.

We get the same message of complexity from the cave art and sculpture that the Cro-Magnons left behind them. Remote parts of caves that were never lived in are sometimes almost saturated with red or black images of the animals that the artists saw in the outside world or in their imaginations – horse, mammoth, woolly rhino, bison, aurochs, deer, reindeer, and more rarely carnivores such as lion, bear, hyaena, and humans themselves. Rarer still there are birds, strange symbols, and weird figures that are part animal, part human. This tradition of cave art lasted over 20,000 years, and it was probably produced for many different reasons, but some of it certainly seems to be linked with spirituality and perhaps ceremonies carried out by torchlight or oil lamplight deep in the caves. There is a parallel tradition of human representation in the form of statuettes, usually female, and often generously proportioned (the so-called Venus statuettes). These figurines may be engraved, or carved from stone, bone, antler or ivory. In some cases (as at Dolní Věstonice), they were moulded from a mixture of clay and ash, fired to a temperature high enough (had the Cro-Magnons wanted) to produce the first pottery. The statuettes are not usually found deep in caves but instead are in living sites, which some people have suggested means they had a quite different significance to the cave art, perhaps special to women. As well as the representational art there are also examples of notched or patterned engraved bones, which have been interpreted as tallies, or calendars based on menstrual or lunar cycles (there is a nice example of these from Gough's Cave in Cheddar Gorge, about 14,000 years old).

ABOVE: This small female figurine from Dolní Věstonice was made from baked clay and bone powder.

BELOW: This mammoth, engraved on part of a tusk, was found at La Madeleine in 1864 and was one of the first relics to show that the Cro-Magnons were indeed contemporaries of the mammoth in France.

All of these behaviours seem to have appeared in Europe after the Cro-Magnons arrived, and we have no evidence that the Neanderthals produced anything as complex. But the Neanderthals did bury their dead, with some possible examples of associated objects such as food (perhaps offerings?) or stone slabs; they knew about pigments, even if they did not paint caves; and they clearly had skills with material such as wood that have almost all perished. We need to bear this in mind when evaluating the relative abilities and complexities of the two populations. However, they clearly differed in their reach across the landscape. Whereas virtually all Neanderthal stone tools were made from raw materials sourced within an hour's walk from their sites, Cro-Magnons were either much more mobile or had exchange networks for resources covering hundreds of miles – for example, amber from the Baltic and shells from the Atlantic or the Mediterranean traversed much of Europe in Cro-Magnon times. This suggests that Neanderthal groups were small, with restricted home ranges, and limited networks of exchange and contact. In contrast, it seems that Cro-Magnon groups were larger and were able to exploit the choicest resources of a region before moving on, establishing and maintaining relationships with neighbouring Cro-Magnon groups as they did so, including networks of trade and kinship.

As we have seen, the Cro-Magnons first appeared in Europe about 40,000 years ago – but where did they come from? The oldest well-dated early modern fossils in Europe are from the cave site of Oase in the Carpathian Mountains of Romania, a location so remote that it requires diving through a flooded and murky 25-metre tunnel to get to it. The skull of an adolescent and the jawbone of an adult found there are the biggest-toothed of all Cro-Magnons, and there are other features reminiscent of African fossils, as well as a detail in the jaw that is found in Neanderthals. Unfortunately there are no artefacts with the Oase remains, which seem to have been washed to where they were found, but early Upper Palaeolithic artefacts from elsewhere in Europe seem to have dual origins. Some seem to develop out of local precursors and this includes leaf points, such as those found in Britain, and an industry called the Châtelperronian (named after the French cave Châtelperron) that seems to have evolved from the French Mousterian of Acheulian Tradition. The only diagnostic fossils definitely associated with the Châtelperronian are those of Neanderthals. But two other early industries, the Bohunician of eastern Europe, and the Aurignacian, found over most of Europe, seem to be intrusive and have links with artefacts found in western Asia and Africa. While no diagnostic fossils have yet been found with the Bohunician, those with the Aurignacian are modern humans similar to the ones associated with the succeeding widespread European industry – the Gravettian, as found at Paviland.

The trail of modern human fossils ultimately leads us back to Africa. There are early moderns associated with pre-Upper Palaeolithic tools in Israel, at the cave sites of Skhul and Qafzeh, including the oldest complex burials known. About 100,000 years ago, a child at Qafzeh was buried under deer antlers, and a man at Skhul was buried clasping the jaws of a large wild boar. Beyond these, we only find modern humans and their ancestors in Africa. From sites in Ethiopia (Omo Kibish and Herto), and Kenya (Guomde) there are rather primitive, but essentially modern, human fossils dating from over 150,000 years ago. Just as Europe records the deep roots of the Neanderthals through more ancient fossils such as Swanscombe, Atapuerca's Sima de los Huesos and Pontnewydd, Africa appears to record the evolution of modern humans through fossils like those from Florisbad (South Africa), Ngaloba (Tanzania) and Jebel Irhoud (Morocco). The two human lines began to diverge and go their separate ways over 400,000

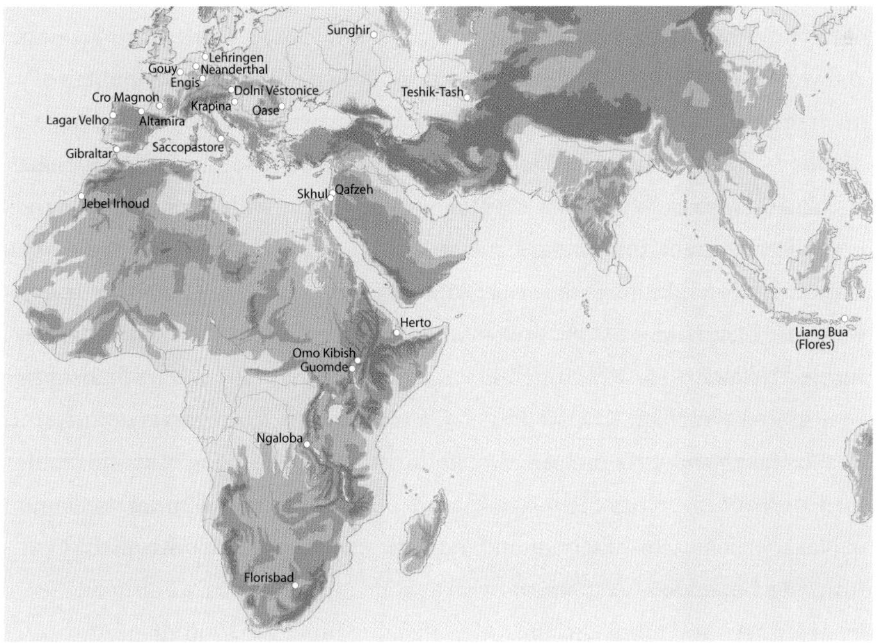

LEFT: A map showing some important Neanderthal sites, Flores, and sites linked to the evolution of modern humans.

years ago, one north of the Mediterranean, the other to the south. Then in the last 100,000 years *H. sapiens* started to emerge from its African homeland, probably first turning eastwards and expanding along the coasts of Arabia and southern Asia towards China and Australia. Those who settled in western Asia or Arabia may have been the source of the moderns who started to infiltrate Neanderthal strongholds in Europe, or they may have derived directly from a second wave from North Africa. Whether these ancestors of the Cro-Magnons were led towards Europe by a relatively warm phase about 50,000 years ago, and consequent movements of game, or whether pressures such as population growth fuelled their migration, the European Neanderthals were suddenly faced with human competition for the first time in their history.

We have seen that they, too, were fully human, with brains as large as ours, and they survived successfully through the challenging and changing climates of Europe for hundreds of thousands of years; as we have seen, they descended from people like those at Swanscombe and Atapuerca 400,000 years ago. So how close to us were they? They were not our ancestors, but were they just a special ice age race of modern humans, an extension of the degree of difference we can find in *H. sapiens* across the world today? Thirty years ago, the American anthropologist Bill Howells and I compared skull measurements of Neanderthals with those of early modern humans such as the Cro-Magnons, and with the variation we found between modern people today. In the case of my research, I compared the statistical distance between four regional groups of today – Eskimo, Zulu, Tasmanian and Norse – with those for Neanderthals and Cro-Magnons. While the Cro-Magnons did fit the pattern of being a variant of modern humans, the Neanderthals were far more distinct and just as different from the Cro-Magnons as they were from other modern 'races'. Howells' results were very comparable. At the time neither of us went so far as to argue that this showed Neanderthals were different enough to resurrect the species name they were given by William King in 1864 – *Homo neanderthalensis* – but research since then certainly supports such an idea. In the most wide-ranging of recent studies, the Greek anthropologist Katerina Harvati not only compared the skull shapes of Neanderthals, Cro-Magnons and modern people using sophisticated digital scanning, but added comparisons with a range of skulls from our primate relatives, apes and monkeys. She found that, judging from skull shape, the

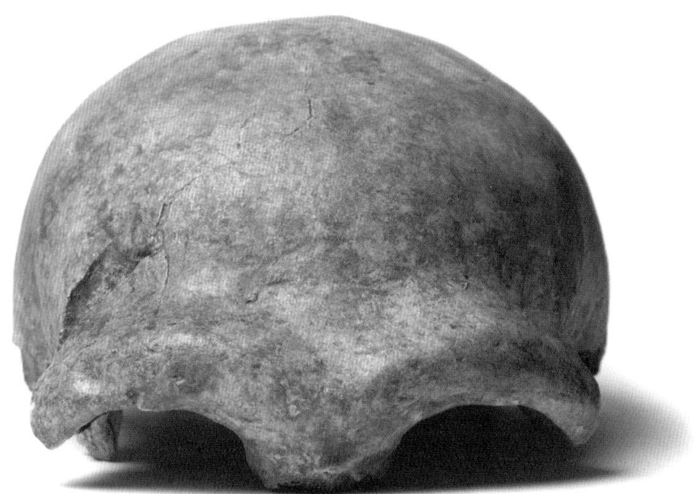

ABOVE: The skull cap found in the Neander Valley (Neander thal) in 1856 – part of the skeleton that gave this ancient population its name.

difference between Neanderthals and modern humans was definitely at the level found today between distinct but closely related species of apes and monkeys.

We do have another source of data about the past, from living people, and even from fossil ones – DNA (Deoxyribonucleic acid). This is the material that stores the complex blueprints (genes) from which our bodies and the chemicals that run them are created. Our DNA is inherited about equally from both our parents, via combinations of the DNA in the paternal sperm and the maternal egg, but there are exceptions to this. Males inherit a Y-chromosome and its DNA only from their fathers, while we all inherit a special kind of DNA from our mothers – mitochondrial DNA (mtDNA) that derives from separate little structures called mitochondria, the power stations of our cells. Analysing DNA patterns in living people enables us to reconstruct how their DNA has evolved and how they are related to each other, since when DNA is cloned from generation to generation, there are sometimes copying mistakes which, if they persist, can be used as markers to track lines of ancestry and descent. The patterns from modern human mtDNA and Y-chromosome DNA agree in placing our shared ancestry firmly in Africa and in the recent past (less than 200,000 years), in good agreement with the fossil evidence. The rest of the DNA that codes for things like proteins, enzymes and body structure also predominantly shows an African origin, but in some cases there are signs of non-African origins, where a DNA pattern seems to be more ancient in, say, eastern Asia than it is in Africa. We do not yet know whether this indicates a measure of modern human blending with more ancient human lines (such as the Neanderthals) outside of Africa, or whether this is an artefact of limited sampling, which will disappear when larger studies are conducted.

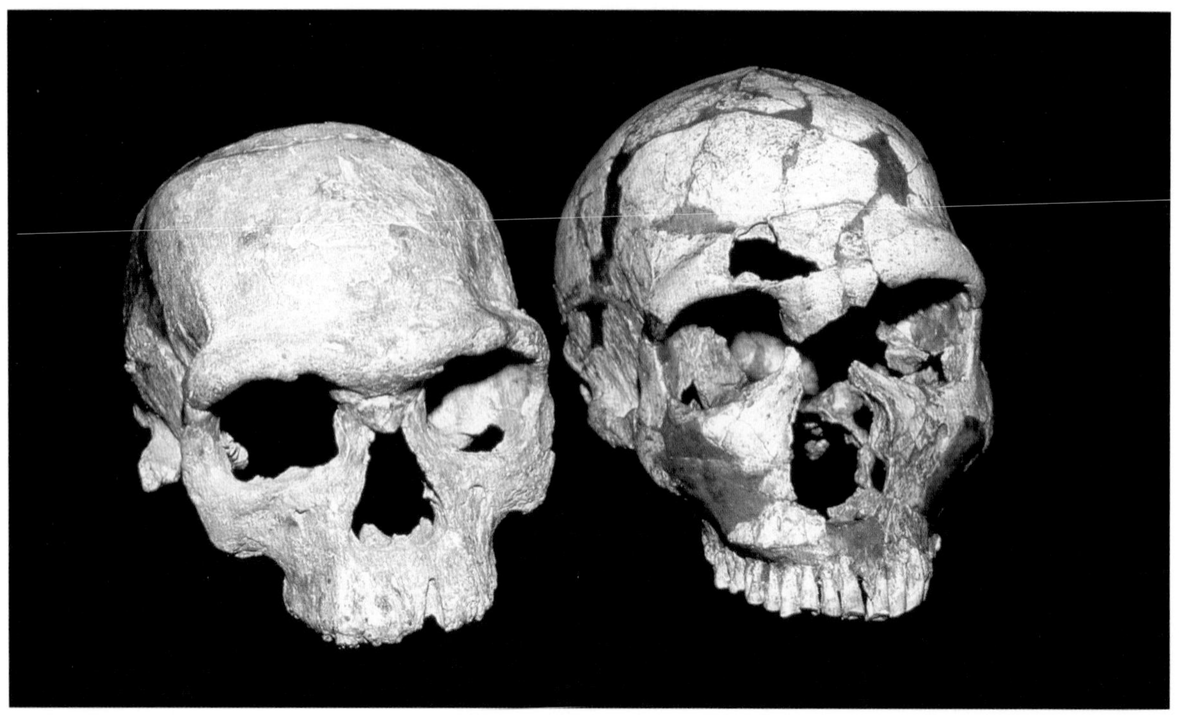

ABOVE: The Jebel Irhoud skull from Morocco (*left*) may represent an ancestor for our species. Despite being more ancient than the La Ferrassie Neanderthal (*right*), it has a facial shape more like that of modern humans.

In the case of mtDNA, a major breakthrough was achieved in 1997 when parts of its pattern were successfully recovered from the original 1856 skeleton from the Neander Valley in Germany, and since then several more Neanderthals have been similarly sequenced. As with the skulls, the mtDNA of Neanderthals is distinct and equidistant from those of modern regional populations, and there is no sign of a closer relationship to recent Europeans, as might have been expected if there had been evolutionary continuity in Europe. The ten Neanderthal fossils that have yielded mtDNA sequences so far reveal differences from each other comparable with those found between modern regional populations, so the Neanderthals seem to have had a population history at least as long and complex as our own. We can calibrate the level of difference between Neanderthals and us, either against our own variation today, or against the differences we show compared with our nearest living relatives, the common chimpanzee and the bonobo. In both cases, the DNA divergence date between us and Neanderthals is estimated at about half a million years, very much in line with a split at the time of *Homo heidelbergensis* and the subsequent appearance of distinctive Neanderthal features in European fossils such as Swanscombe and Atapuerca (Pit of the Bones) dated to about 400,000 years ago. But is that level of divergence enough to denote a species separation? That is not so clear, since some closely related mammal species show that level of DNA difference, while in others (such as the common chimpanzee) such a level of differentiation is contained within a single species. So Neanderthals are in that grey area, either a closely related sister species (as the fossils indicate), or a very divergent lineage of our own species (as the mitochondrial DNA might suggest).

And what about contact and possible interbreeding between Neanderthals and the contemporaneous Cro-Magnons? The mtDNA evidence provides no hint of this so far, and even the species question cannot resolve it, since closely related mammal species may still be interfertile, although the hybrid offspring may not reproduce well, or at all. And the fundamental question is surely whether, when these groups encountered each other, they would have primarily seen each other as essentially 'same' or 'other' – potential friend, enemy, prey even? In my view, after an evolutionary separation much longer than any between living peoples, they would not only have looked very different in their bodies and faces, but probably also in their skin colour, eyes, hair, body hair, gestures, body language and communication. To which could be added whatever clothing or body embellishments were in use as practicalities or social signals amongst the groups of the time. So even if the two species were reproductively compatible, the two populations may hardly ever have wanted to mate if they didn't like the look of each other. And if offspring did result, they may not have been favoured as mates by the next generations of either parent group. So paradoxically, hybrids could have been produced, but their genes may never have impacted the gene pools of the main populations.

BELOW: Reconstructed busts by Maurice Wilson of (*from left*) *Australopithecus africanus*, *Homo erectus* (Java), *Homo erectus* (China), *Homo heidelbergensis* (Broken Hill), *Homo neanderthalensis*, *Homo sapiens* (Cro-Magnon).

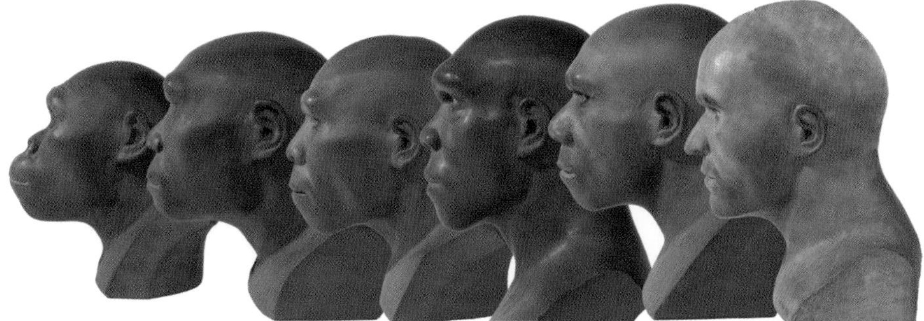

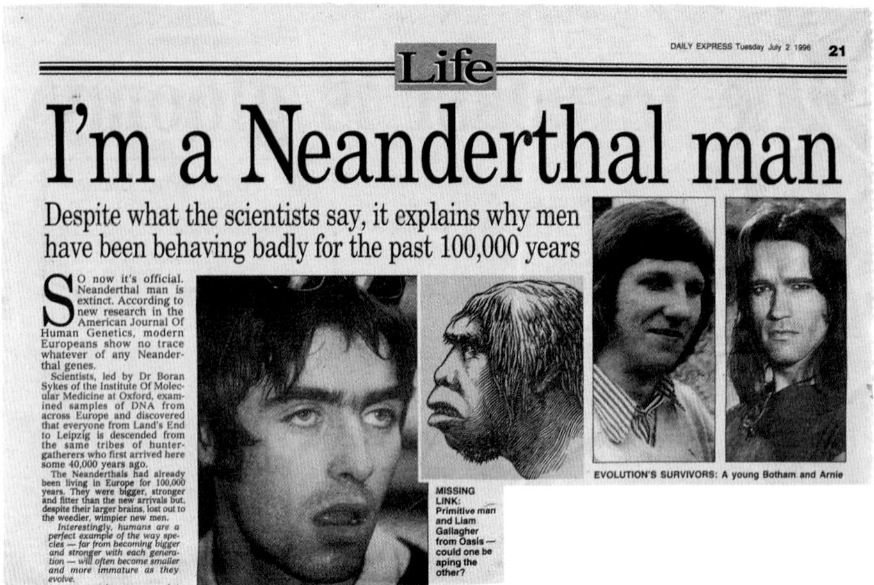

DAILY EXPRESS Tuesday July 2 1996 21

Life

I'm a Neanderthal man

Despite what the scientists say, it explains why men have been behaving badly for the past 100,000 years

SO now it's official. Neanderthal man is extinct. According to new research in the American Journal Of Human Genetics, modern Europeans show no trace whatever of any Neanderthal genes.

Scientists, led by Dr Boran Sykes of the Institute Of Molecular Medicine at Oxford, examined samples of DNA from across Europe and discovered that everyone from Land's End to Leipzig is descended from the same tribes of hunter-gatherers who first arrived here some 40,000 years ago.

The Neanderthals had already been living in Europe for 100,000 years. They were bigger, stronger and fitter than the new arrivals but, despite their larger brains, lost out to the weedier, wimpier new men.

Interestingly, humans are a perfect example of the way species — far from becoming bigger and stronger with each generation — will often become smaller and more immature as they evolve.

MISSING LINK: Primitive man and Liam Gallagher from Oasis — could one be aping the other?

EVOLUTION'S SURVIVORS: A young Botham and Arnie

RIGHT: *The Daily Express* expresses its views about the Neanderthals and us in 1996.

We don't, of course, know what language capabilities either the Neanderthals or the Cro-Magnons had, and can only guess at this from what they left for posterity. The Cro-Magnons were as complex as modern hunter-gatherers and foragers in what we can reconstruct of their ways of life, and as well as their campsites and technology they have left behind evidence of their art, symbolism and music (for example bone flutes). For the Neanderthals, there are hints of social complexity in things like their burial of the dead and (apparently late in their history) production of jewellery. But evidence that they buried their dead with flowers and made bone flutes now seems dubious. Claims have been made that the shape and capabilities of the Neanderthal vocal tract could be accurately reconstructed from fossil skulls and jaws, but even the most pessimistic simulations still give the Neanderthals a big enough range of sounds for complex language, provided brain quality and sufficient social complexity were in place for it to develop. So we must keep an open mind on this, although personally I doubt that Neanderthal social complexity had driven the evolution of their languages to anything nearly as elaborate as ours by the time they died out.

So what really happened to the Neanderthals? Some workers think that there was a relatively gradual merging with the incoming Cro-Magnons, meaning that Neanderthals would indeed have contributed something to future generations of Europeans, even modern ones. They see evidence for this in odd features of Cro-Magnon fossils, which they see as remnants of a Neanderthal heritage. In Britain there is one intriguing, but fragmentary, human fossil that falls into the critical period between 30,000 and 40,000 years ago, when these groups could have been in contact or competition. This is a fragment of an adult upper jaw holding some heavily worn teeth from Kent's Cavern in Devon, found in 1926, long after the time of MacEnery and Pengelly. It was originally described as an early modern fossil and associated with Aurignacian tools, but recent AHOB research suggests that it is instead associated with earlier leaf point artefacts at the site, in which case it is a unique find that could potentially tell us who made these enigmatic tools. Were they Neanderthal, modern, or perhaps even a mixed population? We are now subjecting the fossil to a battery of different techniques, including an attempt at DNA extraction, to see whether we can determine its affinities. In another highly controversial case the burial of a child from Lagar Velho in Portugal, dated about the same time as Paviland and showing similar treatment with red ochre, has been argued to show signs of a mixed Neanderthal–Cro-Magnon ancestry since it supposedly shows the compact build of a Neanderthal rather than the more lithe form typical of the early Cro-Magnons. However, in all other respects the skeleton seems resoundingly modern, and I doubt that it is anything other than an unusually stocky Gravettian child.

BELOW: In 1927 Stan Laurel presented his own strange version of life in Neanderthal times in the film *Flying Elephants.*

There is a range of alternative scenarios offered by those who instead believe that the Neanderthals went extinct because of the arrival of the Cro-Magnons. There might have been warfare between the groups, Neanderthals might have been pushed into marginal and less productive environments, they might have been outbred if Cro-Magnons had better infant survival or food procurement strategies, or the moderns might have brought new infectious diseases to which the Neanderthals had no resistance. The truth is, of course, that we don't even know how often these populations actually encountered each other, but we can guess that population density was low – some estimates put the total population of even the Cro-Magnons at only a few tens of thousands right across Europe. And of course it is possible that given the range across which the Neanderthals lived, from Portugal to Uzbekistan, there was no single cause of their demise. Perhaps the reasons for their extinction in Israel were different from those in the Caucasus, and Italy, and Gibraltar, and Britain.

A new possibility has emerged with our increased knowledge of the complex climatic changes of the last 100,000 years. It had previously seemed unlikely that climate was important in Neanderthal extinction, because the peak severity of the last glaciation seemed to come about 10,000 years after the Neanderthals had disappeared, and they had obviously survived the cold before, and perhaps were even adapted to it, physically and behaviourally. Yet we now know from more detailed records in the Greenland ice cap and European lakebeds that the European climate between 45,000 and 12,000 years ago was highly unstable. One of the primary reasons for this instability was the fluctuating state of the Atlantic Ocean, which has (and had) a huge influence on European conditions. For example, the latitude of Britain (that is, its position in relation to the Equator and the North Pole) is about the same as Labrador in Canada. Whereas in Britain today summer temperatures average about 15°C, and winter about 6°C, Labrador may only manage about 9°C and –12°C respectively. The sea off Labrador is infested with pack ice and icebergs for eight months of the year, and snow may remain on the ground for up to eight months.

As touched on before, the reason why Britain has such warm summers and, especially given its position, such mild winters is the action of the Gulf Stream, also known as the Atlantic Drift, or Atlantic Heat Conveyor. This current transports warm subtropical ocean waters from the Caribbean up to north-western Europe. When these waters have finally given up their warmth, they either flow back along the Greenland and Canadian coasts, or sink into the deep Atlantic circulation heading back south. So today, the polar front that marks the junction between arctic and non-arctic waters lies beyond Iceland. But on many occasions in the past 100,000 years, for reasons that are still not fully understood, the Gulf Stream has completely shut down and the conveyor has rapidly swung into reverse, surrounding Britain with the freezing waters its latitudinal position would otherwise dictate. The polar front migrated towards the Equator, often lying as far south as the coast of Portugal, and even feeding icebergs into the Mediterranean. But after a few thousand years, the warm flow was restored again until a further interruption, and these oscillations back and forth had a huge effect on the climate of Britain and western Europe. What is as remarkable as the number of times this climate switch was turned on and off is the rapidity with which it sometimes happened. Layers of compacted snow in the Greenland ice cap have been laid down annually, and these layers give very precise information on the state of the atmosphere (and oceans) at the time. Astonishingly, some of these extreme oscillations happened over only about ten years.

I worked with colleagues in the Stage Three Project at Cambridge (focusing on MIS 3, roughly 60,000–25,000 years ago) to look at detailed climate records in Greenland and Italy and investigate how this instability might have stressed the human populations of Europe (whether Neanderthal or Cro-Magnon). We assessed only low temperatures and rapidly changing temperatures in our investigation (although many other aspects of the weather were undoubtedly fluctuating too), but we showed that the effects peaked in a prolonged period dominated by cold and instability, around 30,000 years ago. We concluded that two factors were predominantly responsible for Neanderthal extinction in western Europe: the arrival of a competing human population, and the unstable climate. Left to themselves, the Neanderthals might have got through the worst of the climatic changes, as they had clearly managed to before, by going extinct locally and surviving in refugium to the south (a 'refugia' is a place of refuge where conditions are not so severe). They would have bounced back whenever things improved. Alternatively, in a stable environment to which they were adapted, they might have been able to compete with the newcomers effectively, and perhaps would still be with us today – an extraordinary thought. But in such unstable times, with severe climate swings happening even within the lifetime of a single Neanderthal or Cro-Magnon, it would have meant survival of the most resourceful and adaptable at a time when environmental change must have been at its most challenging. The Cro-Magnons surely suffered badly too, and we know that even they succumbed in Britain about 25,000 years ago. Yet maybe with the aid of better technology, housing, clothing, infant care and wide-reaching social networks, they somehow got through the bad times in refugia further south, and we find them coming back from the Continent about 15,000 years ago, as the next chapter shows. The unlucky Neanderthals, however, never returned.

CHAPTER SIX

WHAT THEY GORGED IN CHEDDAR

PRECEDING PAGES: The arctic mountain landscapes of the Cairngorms National Park in Scotland probably resemble the Mendip region of Somerset as it looked 15,000 years ago.

As we saw in previous chapters, the Devensian was the last cold stage in Britain, and it reached its peak about 21,000 years ago. The Last Glacial Maximum, as the event is known globally, was a major event in shaping the world as we know it. The growth of huge ice sheets changed the Earth's climate in terms of atmospheric and ocean circulation, rainfall, and a huge drop in temperature, and where the glaciers sculpted the Earth or shed debris or melt waters, they formed the landscapes we know today. Most of the extra ice that formed was in the large ice sheets that descended from the Arctic, of which the Greenland ice sheet is now the only significant remnant, but glaciers also spread on every high mountain range, including those in the South American and African tropics. The Laurentide ice sheet covering north-eastern America was the largest glacier complex in the northern hemisphere, with an astonishing ice volume of some 35 million km3 (8 million cubic miles), and south of it huge lakes of melt water formed, of which the Great Lakes are just the latest example.

By comparison the Scandinavian ice sheet was only about 6 million km^3, but nonetheless was more than 2 kilometres (1.25 miles) thick, 400 kilometres (250 miles) wide and about 300 kilometres (200 miles) long (north to south), extending west into the Norwegian Sea, south across Denmark and the North German plain and Poland, and east into Russia.

In Britain the major ice accumulation ranged from the mountains of West Scotland to the uplands of Wales, and at its peak it reached what is now Swansea, Wolverhampton and Lincoln. Only parts of the Midlands and southern England were totally free of ice, but they still suffered frozen soils and landscapes of dust formed from ground-up rocks, scoured by winds from the front of the ice caps. Even the volume of the small British ice sheet was about 800,000 km^3 (200,000 cubic miles), and it was up to 1.5 kilometres (1 mile) thick. The British and Scandinavian ice sheets may have merged in the north, but an ice-free corridor of cold desert over 100 kilometres (70 miles) wide separated the ice masses where the North Sea is now.

The huge expanses of ice and snow cover accentuated the global temperature drop by reflecting back the sun's rays, and the amount of water locked up in the ice and snow caused a drop of some 125 metres in global sea level compared with today. While the spread of deserts and arid steppes forced many warmth- and moisture-loving species of plants and animals in the middle of continents to retreat into refugia, others migrated across the vast newly exposed shelves of land around the world's continents. These briefly created the largest land bridges of the recent past, allowing many species of plants and animals to migrate between land masses and islands to where they are found today. In Britain, though, the severity of this period seems to have been too much even for the resourceful Cro-Magnons to cope with, and as the north of Britain sank under the effects of an ice cap a kilometre thick and weighing millions of tons, the few remaining people disappeared from the desert-like tundra south of the ice, either following the migrating herds back across the land bridge to the Continent, or dying out as they tried to cling to their homelands.

About 16,000 years ago, the climate of western Europe suddenly took a turn for the better, the ice retreated, and rich herds of game migrated back, followed by the late Ice Age hunters who lived off them. One British site in

ABOVE: Cheddar Gorge was a renowned beauty spot long before its caves were explored and opened to the public.

particular tells us who these people were, and how they briefly thrived during what is known as the Late Glacial Interstadial: Gough's Cave. This show cave lies on the southern side of Cheddar Gorge in Somerset. Once it was probably the richest Palaeolithic site in Britain in terms of the density and number of finds preserved, but sadly only a part of what was discovered survives today. Nevertheless it still has the largest collection of late Upper Palaeolithic material – both artefacts and fossils – from a British cave, and in the last fifteen years a research team who now form some of AHOB's core membership has led a concerted effort to extract the maximum information possible from what survives. This effort has given us a wonderful, if sometimes disturbing, insight into the life and times of Britain's inhabitants near the end of the last Ice Age.

To understand the recent history of Gough's Cave, we need to return to the 1800s, when commercial rivalry in the West Country beauty spot of Cheddar

Gorge led rival proprietors of show caves to try and outdo each other in what their caves had to offer the visiting public, even to the extent of buying in stalagmites from other caves to build fake 'fairy grottoes'. For most of that century Cox's Cave, with its beautiful formations, held sway, until Richard Gough, a nephew of the discoverer of Cox's Cave, purchased the lease on a cave known as the Great Stalactite Cavern. Realizing his site, despite its imposing name, could not match Cox's, he searched for new opportunities in the Gorge. In 1890, fifty yards to the east, he started to open out the entrance of another cave, which until then had been only a minor tourist attraction as well as, at times, a gambling den and a cart shed. The site was known to have produced fossil bones in a visit by the Geologists' Association in 1880, but even in 1899 the report of a visit by the Bath Natural History and Antiquarian Field Club concentrated on the frugality of their lunch rather than what was seen. Nevertheless Gough and his two sons had been very busy, since the *Wells Journal* of 1892 reported that to expose stalactite formations further back in the cave, they had removed some 500 tons of sediment. From this had been collected 'a large quantity of bones and teeth of extinct animals, besides a lot of flint knives and bone instruments, on which he [Gough] sets great value'. Despite this value, by 1898 a further 500 tons had been quarried away. Perhaps the spoil that was wheelbarrowed out was

LEFT: William Gough at the entrance of his cave.

dumped nearby; if it lies somewhere under the present road through the Gorge, the rich finds it must have contained may one day be recovered.

Work has continued to develop Gough's Cave as a tourist attraction right up to the present day, and this has led to many important discoveries, some intentional and well-recorded, others accidental and without context. Almost all have come from the Vestibule, close to the present metal gates just within the entrance of the cave, where daylight would still have reached the hunter-gatherers who camped there. Shortly after Richard Gough died in 1902, interest in the site was heightened by a discovery made during the lowering of the floor to deal with flooding. This was the skeleton of a young adult male known as Cheddar Man, in the location now known as Skeleton Pit or Skeleton Rift. When Cheddar Man was found, there were claims that he was the long-sought earliest Englishman, with exaggerated dates of 40,000 to 80,000 years old, but none of the tools found in the cave give any hint of such an antiquity, and as we will see he lived, in fact, at the beginning of the Holocene, the recent geological phase in which we are living now. Nevertheless, as we will also see, he does have a special British importance.

ABOVE: Excavations in Cheddar Gorge. Tons of sediments containing rich archaeological finds were removed from Gough's Cave in order to allow greater public access to the deeper chambers.

OPPOSITE TOP: Two of Richard Gough's sons pose proudly by the newly discovered remains of Cheddar man.

Those who first studied the flint tools from Gough's realized that they were Upper Palaeolithic and specifically resembled those of the Magdalenian industry of France, which dated from near the end of the last Ice Age. The similarity was enhanced by the discovery of an artefact known as a bâton, made from a perforated reindeer antler, typical of the Magdalenian. The stone tools also resembled those known from other British sites such as Robin Hood Cave at Creswell Crags (Derbyshire) and Kent's Cavern in Devon. This led the eminent British archaeologist Dorothy Garrod to recognize a specific British industry she named Creswellian for the stone tools from these sites.

By 1931, at least 7,000 stone artefacts had been recorded from Gough's (unfortunately even most of these have since been lost), as well as tools made of bone, antler and ivory, including a second bâton made from reindeer antler, and the sharpened shin bone of an arctic hare with grouped notches along its edges. What these were used for can only be guessed at – the hare bone would certainly have been a good skin and fur piercer for making clothes, while the function of the bâtons remains an enigma. They were once assumed to be the emblems of chiefs that hung round their necks or waists (*bâtons de commandement*), but they are spirally grooved and worn by use where they have been perforated by a large round hole. This suggests they could have been used for working leather or fibres, for straightening wood or even bone and ivory (possible if it was steamed or heated). Perhaps they were even used as a pulley or a device for corralling animals, as AHOB member Andy Currant has suggested – imprints of ropes are known from the famous French Magdalenian cave of Lascaux.

As well as these functional artefacts Gough's Cave has, over the years, produced evidence that suggests the site might have had some more special meaning for its inhabitants. These finds include a piece of animal rib that from its appearance must have been kept and handled for many years. On one side it is marked with scores of lines along its edges, like a pocket ruler, while on the

BELOW: This bâton made of reindeer antler was discovered during our excavations at Gough's cave in 1991.

E.1596

other it has many diagonal intersecting lines forming a criss-cross pattern, into which the iron oxide pigment red ochre has been rubbed. There are also pieces of Baltic amber that must have been carried at least from the present area of the North Sea – perhaps once items of jewellery. The presence of the amber so far from its source suggests that either these people were very mobile, or they had long-distance trading networks. Mobility is also implied from study of the flint used for tools at Gough's – it most probably came from Salisbury Plain, some 70 kilometres (45 miles) away.

OPPOSITE: Worked bone and antler from Gough's Cave: three antler batons, a point made from hare bone (*top right*) and the enigmatic 'tally-stick' (*top left*).

BELOW: A 14,000 year old bone needle from France. Not only did the Cro-Magnons have the technology to make sewn clothing and tents, they also knew how to weave and make ropes.

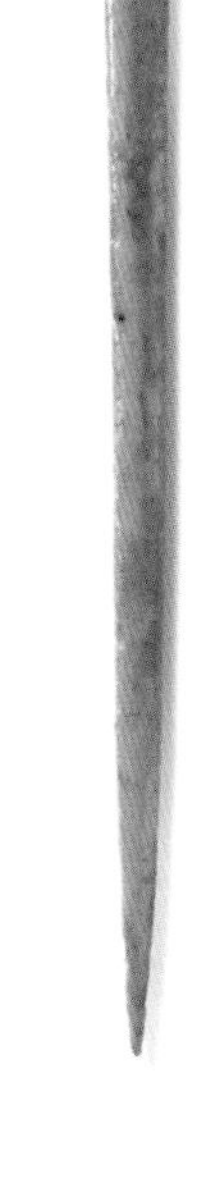

Many of the animal remains from Gough's Cave show signs of modification by humans using flint tools. Some of this is no doubt the accidental by-product of dismembering a carcass, but in other cases such as the working of hare bone points there was certainly an end product in mind. The fossil animal bones help to paint a vivid picture of Somerset near the end of the Ice Age. There were large herbivores such as horse, red deer and the extinct giant ox providing meat on the hoof, while smaller animals such as badger, arctic hare and birds such as black grouse, ptarmigan and partridge were also consumed when available. One of the bird species, whooper swan, was even used as a source of bone, probably for making needles. There are a few other notable species such as the saiga antelope, now found on the arid steppes of central Asia, and two species of lemming, but the expected cold climate species of mammoth and reindeer are in fact only represented by worked material such as bâtons and spear points that could have been carried from elsewhere. Finally, and especially intriguingly, wolf bones had been identified from Gough's. However, AHOB researchers have noted how small these 'wolves' are compared with the large size of ice age wolves and consider this to be the first British evidence of domestication, as these are actually dog bones. European researchers have made similar claims for Magdalenian sites in Germany, where they were apparently used in the hunting of wild horses. There is no doubt that dogs would have been valuable allies in the struggle for survival, and we can envisage a mutually beneficial relationship gradually developing with tolerance between humans and stray wolves, or orphaned cubs being adopted, and then used selectively to breed strains with the desired submissive and cooperative qualities.

The climate at the time Gough's Cave was first occupied can be reconstructed from plant and beetle remains some 15,000 years old, preserved

ABOVE: Cheddar Gorge contains many caves, some of which remain virtually unexplored.

in a small lakebed at Llanilid in South Wales. The beetles were studied by AHOB Associate Russell Coope, and from the present-day preferences of the species concerned average summer temperatures must have been as high as those of Britain today. But the winters were much colder, and it was also much drier. Combining this with the animal species preserved at Cheddar, we can imagine treeless moorland on the top of the Mendip Hills, descending into bushy scrub. Finally there would have been a narrow zone of tree-cover along the southward slopes and in sheltered valleys such as the Gorge, providing wood for fuel and artefacts. Where there was no shelter or surface water, the lowlands would have been grassed, giving excellent pasture for the wild horses that are the most common large mammals in the cave deposits. Because of its location in the landscape and its shape, Gough's Cave was in many ways an ideal home base for bands of hunter-gatherers 15,000 years ago. The River Yeo emerges at the nearby Risings, and, as just mentioned, the cave was well placed for the exploitation of a variety of plant and animal resources. At the time it was occupied, it probably led straight out into the Gorge, providing plenty of opportunities for ambush, or to drive game against the steep walls. Despite the depth of the Gorge, the large and arched entrance would have let in sufficient daylight for people to group and work within the shelter of the cave. Although relatively dry at the mouth, it would have had convenient sources of water further inside for drinking and cooking. The relative coolness of the cave might also have been useful for storing meat during the warm summer months.

As already mentioned, the animal with by far the largest number of surviving bones and teeth at Gough's is horse, and the remains show extensive damage in the form of cut marks to remove meat, and breakage to extract marrow and brains. One particularly characteristic pattern is the fracturing of the teeth of the lower jaw caused when it was split for marrow. Horses were taken at all ages, so it seems the hunters had the pick of their prey. Although the collection of bones and teeth surviving is a biased one, dominated by the parts that would have been noticed during unsystematic collecting, most parts of the skeleton are represented, suggesting that horses were killed near the cave and that the occupants had the whole carcass available to process. Cut marks on the teeth and bones of both horses and red deer come from skinning, butchery, and the stripping of tendons and ligaments – the latter could have been used as thread for fastenings, or even as rope. Even horse hooves seem to have been removed from the hoof-core, perhaps to extract glue from the keratin, so little seems to have been wasted. However, although there are traces of burnt bone, charcoal and heated stones in the cave, it appears that much of the meat was eaten raw.

Over the years of controlled and uncontrolled excavations in the cave, more human bones were recovered at Gough's, including the skulls of an adult and a child, but until recently it was unclear how, if at all, they related to the remains of Cheddar Man found in 1903. However, excavations near the cave entrance led by Roger Jacobi, Andy Currant and myself have thrown considerable light on this question. In 1986 Roger had been checking some crumbling cave sediment, thought to be old spoil, under the wall of the cave near Skeleton Rift, when he found a human tooth. Further investigations produced more teeth, and over the following few years a series of small excavations were carried out. These showed that this small pocket of

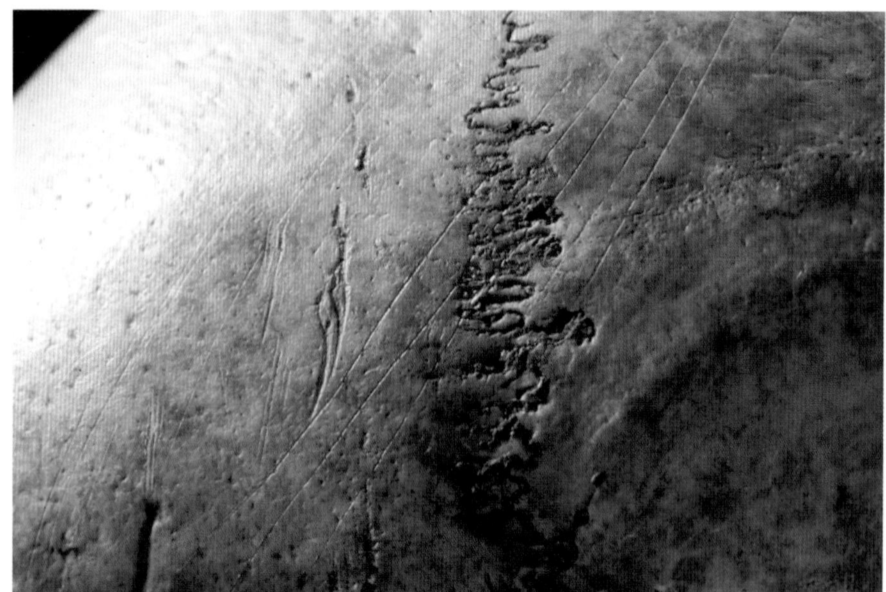

RIGHT: This human skull found in our excavations at Gough's Cave in 1987 has a series of cut marks made by flint tools indicating defleshing and possible scalping of the man concerned.

deposits had been undisturbed for some 14,000 years, and this has led to a whole new chapter in the story of Gough's Cave and its importance to the ancient occupation of Britain.

The excavations also showed how rich a site Gough's Cave must have been before it was decimated from 1890 onwards. One of the first finds were the jawbones of an adolescent boy, into which the teeth that Roger had found fitted perfectly. Subsequently a piece of frontal bone was discovered that refitted on to a more complete adult's skull found some sixty years earlier, showing that this remaining little pocket was indeed the same age as the earlier finds. But the most complete and disturbing of the finds was made by Andy Currant: an adult skull cap on which could be seen a series of cut marks across the side walls and brow. These marks were so well preserved that their direction and angle could be reconstructed to show that the skull had been held in someone's left hand while they cut with a stone tool in their right hand, apparently to scalp the man in question and to cut through muscles holding the lower jaw. On further inspection, it turned out that all of the human skulls and jawbones found in the Creswellian layers (but not Cheddar Man) had been cut with flint knives. What story lay behind these disquieting discoveries?

The management of Cheddar Caves, who had kindly facilitated our work there, naturally wanted to publicize the finds and emphasize the continuing scientific importance of their cave. At the ensuing press conference, we tried to cover the various possible scenarios of defleshing for ritual burial, social consumption of the remains of close relatives, crisis cannibalism (where starving individuals are forced to eat human flesh in order to survive), and cannibalism following violent encounters. The media, of course, only had ears for one version of the story, and a series of lurid headlines soon followed, including 'Stone Age Brits Ate Kids', 'Not So Gorgeous' and 'What They Gorged In Cheddar'. This also led to appearances on a children's breakfast TV show with one of the skulls, probably impacting the sensibilities and appetites of watching parents more than their offspring. However, detailed study of the material has revealed a more complex series of possibilities. My research with a colleague Louise Humphrey suggests that the Creswellian human material from Gough's represents a minimum of five individuals: a young child of about three at death, two adolescents (one probably male), and two adults, one of whom was male. If the human bones and teeth are those of some of the inhabitants of the cave, the presence of a young child suggests that these included families, but if the material was brought in from elsewhere, this is less certain. The individuals concerned were large-bodied, muscular and big-toothed by modern European

BELOW: The evidence of possible cannibalism at Gough's cave led to a series of lurid headlines.

THE DAILY TELEGRAPH, WEDNESDAY, JULY 29, 198

'Cannibal evidence in Stone Age dig

By Paul Stokes, West Country Correspondent

IN WHAT is said to be the most important find for its period in northern Europe, archeologists sifting the remnants of a Stone-Age "dustbin" believe they have found evidence of cannibalism among our ancestors.

It is also described as "the richest pocket of artefacts relating to human occupation ever recorded in Britain."

A second child and at least three adults have now been accounted for.

The human bones and obvious food bones were all cut in the same way and mixed together on the same site.

Preliminary findings show the 12-year-old boy had been both defleshed and detongued while the other skeletons had been defleshed and one skull showed signs of scraping in the eye sockets.

Other insights are given into the eating habits of Stone Age man with the discovery of the remains of beavers, horses, deer, hares and the Saiger antelope.

Skeleton

Germany Japan

gorge

Hazards of meal in

captain Simon Hayward told a how he became the innocent national drugs ring.

From RONALD RICKETTS in Stockholm

Ancient Brits ate children

...aled: The ...ibals of Cheddar ...ge . . .

...HIRSTY discovery about Stone Age man ...ed as the 'find of the century' by natural ...erts.

...bones found by archaeologists excavating around the ...ge in Somerset show ancient Britons ate each other — ...eir own children.

...discovery is the first hard evidence of cannibalism

Daily Mail Reporter

What they gorged in Cheddar

THE discovery of remains from a cannibals' feast held 12,000 years ago was hailed today as the ... of its type in

had been hacked with flint knives and axes. Marks on the skulls suggested the victims' tongues and eyes had been cut out.

The bones were cooked ... dumped on a pile

Victorians blasted the entrance of Gough's Cave.

One of the skulls is almost complete and fragments from another match parts of bone found 70 years ago.

The find was made during a ... by experts

standards, but their teeth were relatively unworn and in good condition, with no signs or markers of stress during growth, or disease.

Study of damage to the bones by archaeologist Jill Cook, and by AHOB Associate Peter Andrews and his Spanish colleague Yolanda Fernández-Jalvo, showed that most of the limb bones were fragmentary and had been deliberately smashed, the rib cages had been opened up, and damage to a neck vertebra even suggested that one individual had been beheaded while lying face down. In deciding whether cannibalism had occurred in such an assemblage, Peter and Yolanda examined four main criteria. Allowing for anatomical differences, were animal bones (assumed to represent food debris) and human bones treated the same in terms of the methods of butchery? Were there similar patterns of breakage in the long bones that might have facilitated marrow extraction? Were there similar patterns of post-processing discard? And was there evidence of cooking, and, if so, was it applied equally to both animal and human remains? Because of bias in the way the collections were made (the animal bones were mostly gathered unsystematically and almost randomly, while the human remains mostly consisted of parts of the head and jaws that would have been easily recognizable, as well as ones we had excavated carefully), examination of the first three criteria was difficult. Nevertheless, comparing the treatment

REMARKABLE DISCOVERY IN CHEDDAR CAVES.

Mr H. St. George Gray, the curator of the Taunton Museum, who was formerly assistant to the late General Pitt-Rivers, the celebrated archæologist, has just examined, at the request of the Somerset Archæological and Natural History Society, a remarkable human skeleton which has been discovered in Gough's Caves at Cheddar. The remains are believed by Mr Gray to be those of a cave-dweller who lived between the Palæolithic and the Neolithic ages. The skull is in many fragments, and encrusted in loam. The man had very prominent brows. The forehead is of the usual width, but, on the other hand, it is very receding. The lower jaw is powerfully formed, and far beyond the width of those of the present age. The skull is also very thick. The thickest part of the frontal bone is nine millimètres, while the average of the present day is only seven millimètres. The skeleton was found between two layers of stalagmite. The height of the man was 5 feet 3½ inches. In the Stone age the average height of a man was only about 5 feet 3 inches or 5 feet 4 inches, and that of a woman 4 feet 11 inches. The shin bone is flat, which is never the case at the present day. Flint implements were found near the skeleton of the type used by cave men.

RIGHT: When Cheddar Man was discovered he was promoted as one of the most ancient human relics.

> The rib cages had been opened up, and damage to a neck vertebra even suggested that one individual had been beheaded while lying face down

and disposal of the most common large mammal remains (horse and deer) showed patterns comparable to each other and to the human remains, thus supporting the case for nutritional cannibalism. This even extended to details of the dismemberment of the head from the jaws, of tongue removal, and jaw breakage to extract marrow (although for the fourth criterion, there is little evidence of cooking on any of the bones).

Jill Cook of the British Museum has instead suggested that many of the marks and damage on the human bones were produced by trampling as the bones lay on the cave floor, and she has questioned whether cannibalism would ever have been necessary on nutritional grounds. Instead, if it happened at all, it could have taken place as part of a funerary ritual. Some peoples today dismember and lay out bodies on platforms, to be eaten by vultures, while others dig up the corpses of loved ones, strip off the flesh, and rebury them. Roger Jacobi accepts the reality of the processing marks on the human bones, but thinks that these procedures might have been carried out to produce manageable packaged transport of bodies to the cave, perhaps in bags, if group members had died away from their home base. One surprising result of the radiocarbon dating of the animal and human bones found in the surviving pocket of sediment next to the cave wall is the wide span of the dates, covering over a thousand years. This might suggest that the accumulation represented a redeposition, perhaps as part of a final act of site cleaning before the cave was eventually abandoned. In which case, both the animal and human accumulations resulted from repeated episodes of human occupation, accompanied by disposal of both the animal and human dead.

As we saw in the last chapter, characteristic levels of carbon and nitrogen isotopes in different foods we eat (plants, meat, fish) are taken up into our bones, and hence carry signatures of our diet. Isotope analyses carried out on some of the Gough's Cave Creswellian human bones by a team led by AHOB member Mike Richards are particularly interesting. These revealed that the people concerned had consumed mainly animal protein, consistent with

OVERLEAF: The prairies of the Mendips would have been covered with herds of wild horses like these about 14,000 years ago, and they formed the main source of meat for the hunters and gatherers who camped at Gough's Cave, in Cheddar Gorge.

ABOVE: This bone spear-thrower from Bruniquel Cave, France, is about the same age as the Creswellian finds from Gough's Cave. It is beautifully carved with depictions of reindeer.

hunting and carnivory. While the isotope signature indicated that the Cheddar people were certainly high up the meat-eating food chain compared with a contemporaneous arctic fox also sampled, the signal expected from the consumption of grass-feeding horses was not present, and suggested instead that deer, giant ox or perhaps even other carnivores were the regular prey of these people. Interpretation of the results is compounded by uncertainty as to whether the human remains were of the cave residents, thought to have been primarily horse-hunters, or from a different group of humans brought in from elsewhere to be processed and presumably eaten. And there is one extra complication: if human flesh was regularly eaten, it might have overprinted an isotope signature from the consumption of horsemeat.

At the time of the Creswellians, sea level was lowered over 75 metres and Britain was joined to Europe by a wide land bridge, which would have included the rich plains and hills of Doggerland now lying under the North Sea. The nets of trawlers have been dredging up the bones of Ice Age mammals such as mammoth, woolly rhinoceros, horse, deer and reindeer from 20–50 metres deep for at least 200 years, and occasionally flint, bone and antler tools are also found. When Gough's Cave was occupied, similar peoples were living in Holland, Belgium, Germany and France, but the northern lands were always more sparsely settled and human groups probably had to be very mobile to make the best of scattered resources and migrating game. The particular distinguishing features of Creswellian flint work include a lack of very small tools and the dominance of particular kinds of blades, blunted along one edge and often abruptly chipped across the ends – thought to be spear points. The closest parallels with those on the Continent are from sites in Holland and Belgium and it may well be that these were actually the same people migrating seasonally or through their lifetimes to and fro across Doggerland to the Midlands and West Country. This possibility is reinforced by the amber present in Gough's, which could have come from the Baltic or from glacial erratic material pushed southwards by ice into the hills of Doggerland. The fact that the similarities lie to the east rather than the south suggests that the Creswellian colonizers of Britain descended from Magdalenians who had already moved north and responded to a rapid climatic warming that marks the beginning of the Late Glacial Interstadial about 16,000 years ago.

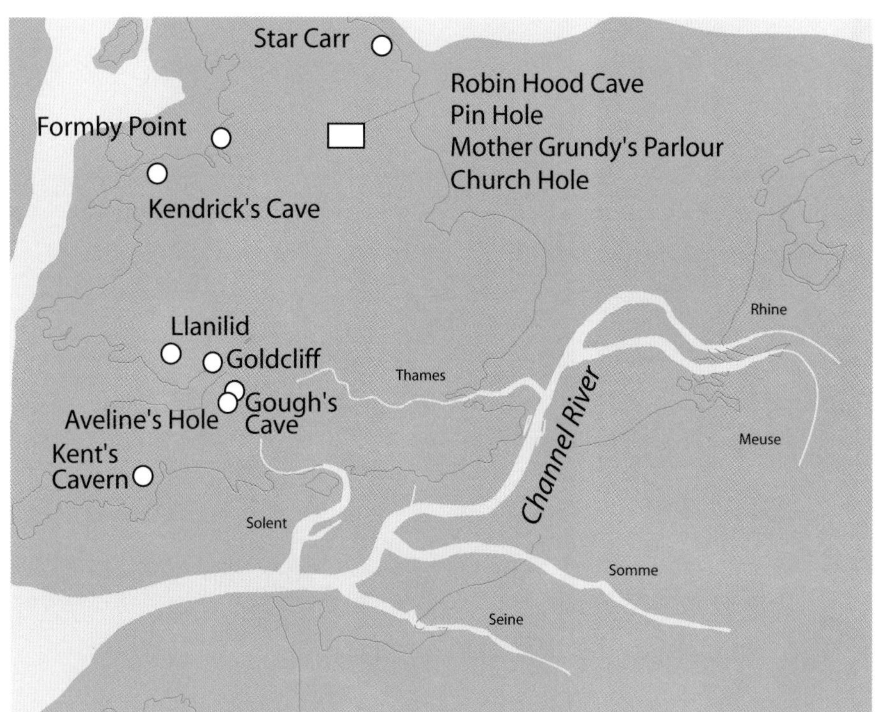

LEFT: A reconstruction of Britain as it may have looked at the time Creswellian people occupied caves in Creswell Crags and Cheddar. Some later (Mesolithic) sites are also shown. The extensive Channel River system may have forced migrating populations further North, across Doggerland.

Another significant difference between the European Magdalenian sites and their Creswellian equivalents, such as Gough's Cave, has been the lack of figurative art in the British sites, either in the form of portable engraved or sculpted pieces, or art on cave walls. The only definite examples known were the engravings on rib bones of a human figure from Pin Hole Cave, and a horse from Robin Hood Cave, both in Creswell Crags (a limestone gorge running east to west and forming part of the border between Derbyshire and Nottinghamshire). It was assumed that this difference reflected the relatively impoverished and challenging conditions in which the early Britons had to live, or the greater sophistication of their continental neighbours. But in 2003, there was a major breakthrough in our knowledge of British cave art, giving a more sensitive and human dimension to the Creswellians than the images of possible cannibalism at Gough's Cave. Three Palaeolithic archaeologists, two British (Paul Pettitt and Paul Bahn) and one Spanish (Sergio Ripoll), considered that such art might exist somewhere on a British cave wall, and could have been missed in previous inspections. With the benefit of their previous experience in locating and studying cave art, they began a careful survey, fortuitously starting at Creswell Crags. On their first morning they noticed some apparently

ABOVE LEFT: An incomplete Creswellian engraving of a bovid (bison or wild cattle) dating from around 13–14,000 years ago, from Church Hole, Creswell.

ABOVE RIGHT: The view looking out from Robin Hood Cave.

non-figurative marks in Robin Hood Cave, and other marks, including what looked like an animal's head, in Mother Grundy's Parlour. Next they visited the unpromising locality of Church Hole, one of the less famous of the Creswell caves. With the benefit of highly directional lighting, they scoured the cave walls and to their astonishment, despite the distraction of recent graffiti, soon recognized the engraved shape of a large animal, about 45 centimetres (20 in) long – apparently an ibex.

Pettitt, Bahn and Ripoll believe they have now identified some ninety possible engravings on the walls of Church Hole, many of them just patterns of lines, but including depictions of horse, bison, bear, birds and women. Further study suggests that the first engraving identified is actually that of a stag, not an ibex, which fits better with our knowledge of the British Ice Age fauna of the time, in which ibex is unknown. As with much of the cave art on the Continent, it seems that natural undulations in the cave wall suggested the shape of a particular animal to the artist, who then amplified this by engraving ears, eyes or muzzles. But there are also stylistic peculiarities in the art, since most of the representations comprise only parts of the animal, primarily the head or forequarters, and two bison heads are so similar that they were surely produced by the same hand. As with all remarkable discoveries in archaeology, it is always wise to consider alternative explanations such as, in this case, more recent forgery. But the style of the art and the species represented are certainly consistent with Magdalenian examples from Europe, and final proof that the art was ancient has come from uranium-series dating of cave flowstone covering parts of the stag, by Alistair Pike of Bristol University, giving an age of over 13,000 years. The Church Hole finds are significant for many reasons, beyond their age and what they depict. First, they suggest that careful searches of other British sites may reveal similar, or perhaps even more spectacular, examples of cave art. Second, they show commonality with European finds to the south and east, suggesting that human groups at this time were indeed linked across what is now the North Sea and the Channel. Third, this country has at last been added to the distribution map of decorated Ice Age sites in Eurasia that stretches from Portugal to the Urals. The most northerly known decorated cave has moved some 450 kilometres (280 miles) from Gouy, near Rouen, to the British Midlands, and the sparse portable objects at Creswell have at last been richly augmented.

Summer temperatures in the early part of the Late Glacial Interstadial were as high as those in Britain today, but it was not to last, for we know from plant and beetle remains that within two thousand years the climate had got wetter, but the temperature had dropped by at least 4°C. These changes seem to have led to the spread of birch forest in Britain, and this in turn would have forced horses to migrate to more open steppe. Evidence for human occupation soon peters out at Gough's with the disappearance of their favourite prey, and we see a change throughout England in the animals and the human occupation patterns. In the latter stages of the Late Glacial Interstadial the large mammal fauna consisted of woodland animals such as red deer, roe deer and elk, but there are also reindeer and wild cattle. And while at the time of Gough's two thirds of known Creswellian sites were in caves compared with settlements in the open, in the latter part of the Late Glacial Interstadial that ratio was reversed. With the arrival of the Late Glacial Interstadial, it looked as if the Earth was at last emerging from the chill of the Last Glacial Maximum into an interglacial,

but the drop in temperature that the last of the Creswellians began to suffer was to get even worse. Soon after 13,000 years ago, the Earth plunged into a final and severe cold snap known as the Younger Dryas stadial, named after the alpine/tundra wildflower the Mountain Aven or *Dryas octopetala* that characterized this period in Scandinavia (as the Latin name suggests, this shrub has a flower with eight petals). To judge from annual layers in the Greenland ice cap, the mean annual temperature dropped by about 15°C in both Greenland and Britain in some ten years, and this big chill was to last over a thousand years. It is believed that the warm conveyor belt of the Atlantic Ocean shut down completely for the last time in human history, but the effects on climate of this chill were global, not only causing the return of ice caps to Scotland and the Lake District, but affecting regions as far away as South America, the Antarctic and New Zealand. While its effects on the hunter-gatherers of Britain were severe or even catastrophic (it is unclear whether any survived here), in terms of human history there were more positive effects, as it is possible that these changes also catalysed the populations of the Middle East and the Far East to begin the domestication of plants and animals.

OPPOSITE: Creswellian artefacts from Gough's Cave and a bone harpoon from Kent's Cavern.

About 11,500 years ago, the Younger Dryas ended as abruptly as it began. Subtropical waters once again fed the Gulf Stream, and northern Europe rapidly thawed out. Forests quickly regained the ground that they had lost to the cold and aridity, and within two thousand years most of the ice sheets were gone. From one extreme to the other, the Earth's climate became even warmer and moister than today for several thousand years. In the north, forests grew closer to the Pole than they do now, while further south, much of the Saharan desert was wet and verdant, with hippos and crocodiles swimming where now there are only vast sand dunes. This phase, known as the Holocene Optimum, occurred between about 9,000 and 5,000 years ago. The hunter-gatherers of Europe who had adapted to the relatively open conditions at the end of the Ice Age now faced new challenges. Animals such as mammoth, cave bear, spotted hyaena and lion, part of the European scene for hundreds of thousands of years, died out locally or globally as dense forests spread across much of western Europe, accompanied by forest-loving animals such as red deer, brown bear and wild boar. As people adapted to the new conditions, they changed their technology, and this marks the transition from the Upper Palaeolithic (for example the Creswellians) to the

Mesolithic (Middle Stone Age), with a tool kit of small stone tools – microliths – many of which must have been mounted on wooden handles, and the spread of the bow and arrow.

A site that illustrates how well early Mesolithic people in Britain managed this transition is Star Carr near Scarborough in Yorkshire. As the last glacial ice retreated, it created lakes of melt water in many places, and one of these was in what is now the Vale of Pickering. The former Lake Pickering is now carr (waterlogged soil with reeds and alders) and its thick deposits of silts and peats have been under excavation on and off for some fifty years. Sealed by peats, an occupation surface with dense concentrations of Mesolithic artefacts of both stone and wood and associated bones has been wonderfully preserved, dating from about 10,500 years ago. The wet conditions conserved many wooden tools, and nearly 200 barbed points of bone and antler, as well as jewellery in the form of stone and amber beads, pierced deer teeth, and thousands of small flint tools. These were dominated by chisel-like burins, presumably for working the considerable quantities of bone and antler found there. At the edge of the ancient lake, amongst reeds, a 6-metre long wooden platform of worked timbers had been built, the oldest such structure known. The individual timbers were skilfully split, from trunks of either poplar or aspen, to form planks up to 3 metres long and about 3 centimetres thick. The only technology available for this in the Mesolithic would have been stone axes and hammers, and stone, antler or wooden wedges. Studies of the many charcoal and burnt wood fragments at the lake-edge suggested that the occupants repeatedly burnt the reed swamp zone to keep their route to the lake clear or to make hunting easier.

ABOVE: The people of Star Carr made beautiful harpoons of red deer antler.

During the Star Carr excavations, extraordinary finds were made: twenty-one worked red deer skulls still with the stumps of their antlers, and all with a pair of holes through the back of them. These must have been headdresses, tied through with a leather thong, and were originally thought to have been worn by hunters as a disguise. The wooden platform was thought to be a mooring for boats, an interpretation supported by the discovery of what seemed to be a wooden paddle. But despite the rich finds from Star Carr, there was little evidence of fishhooks or, indeed, of fish bones. So an alternative explanation for the platform is that it was actually a walkway for depositing offerings in the lake as part of religious ceremonies. In this interpretation, Star Carr was not only an important 10,500-year-old Mesolithic occupation site, but also a ritual centre for people who held the lake sacred. The antler headdresses were probably not worn during hunting, but instead were part of a costume for ceremonies, or perhaps were even offerings thrown into the lake.

ABOVE: Amber, transported from the Baltic or found in deposits from the previous ice advance, was made into beads at Star Carr about 10,000 years ago.

No human bones were found at Star Carr, but if we return to the Cheddar region, finds from there give us a picture of Britons during the Mesolithic era about 10,000 years ago. The skeleton of Cheddar Man, mentioned already, was found during blasting in the entrance of Gough's Cave around Christmas 1903. Early press coverage and reports greatly exaggerated the age of the find, and his death was variously attributed to starvation, rock fall or drowning; it was also said that his diet was probably roots and grains and that he was incapable of walking upright. In fact he appears to be a rare example of a Mesolithic burial, and recent studies suggest that he was a young man in his early twenties when he died. Like his predecessors in the Creswellian, his skull was long-headed and had strong muscle markings, but his brow ridge, jaws and teeth were smaller than in the earlier finds. Like the Creswellians, his teeth were relatively unworn and in good condition, without decay or indicators of stress during growth, but he had his wisdom teeth. This is in contrast to two of the three Creswellian jawbones, which show an unusual feature most common in oriental people of today – they lack their third molars. Such a feature is thought to be an evolutionary response to dental overcrowding, as jawbones have got smaller. Teenagers occasionally do not develop any wisdom teeth, and in a situation where dental overcrowding could lead to dental infections and death, those individuals would have been somewhat more likely to survive, and pass on that feature to their children.
In contrast to the muscularity of his skull, Cheddar Man's body was lightly muscled and his pelvis was rather feminine in shape. He was not tall, at about 166 cm (5ft 5in), but the lower parts of his arms and legs were relatively long, as in the early Cro-Magnons, proportions thought by some anthropologists to be a hint of warm climate evolutionary origins. Cheddar Man appears healthy from his skeleton, apart from one noticeable feature: on his right forehead there is an oval pit that marks the site of a bone infection, perhaps the result of an injury, or perhaps originating from his sinuses. This infection would have formed an open wound, oozing pus, and would certainly have been debilitating, perhaps linked with fever and general malaise, but it is not known whether it killed him.

ABOVE: This jaw bone from Gough's Cave has a series of cut marks behind the molar teeth, indicating where muscles were removed to separate the jaw from the skull.

A cave 5 kilometres (3 miles) from Gough's, in the Mendip Hills of Somerset, has given us further information about the Mesolithic inhabitants of Britain. Aveline's Hole has recently been revealed as the earliest dated cemetery in Britain, used for burials over a period of several hundred years, and as a rare decorated cave, with engraved patterned lines on its walls. Before its discovery in 1797 the cave appears to have been sealed for some 10,000 years, and when opened there were more than seventy skeletons lying 'promiscuously' on the cave floor. Somewhat more controlled excavations in the early twentieth century found flint tools, animal bones and the fragmentary remains of twenty-one more individuals, including what appeared to be a ceremonial burial on a hearth, with red ochre, deer teeth used for necklaces, and even some fossil ammonites. Sadly, a bomb that fell on Bristol in 1941 destroyed most of this remarkable collection together with the excavation records, but the surviving bone fragments indicated that there were adults, young children and two infants. As with Cheddar Man, these people were small and slightly built, but in contrast to the Gough's finds, there were signs of growth stress in the teeth and skeletons, indicating childhood illness or poor nutrition. One adult arm fragment also showed the kind of distortion that comes from repetitive strong activities such as paddling a canoe, or throwing missiles in hunting or fighting. The animal bones in the cave indicated that red deer, wild boar, wolf, lynx and bear may have been hunted for food or their pelts. Isotope analyses of the surviving bones and teeth suggest that the Aveline's people were indeed meat eaters, and as with the archaeological evidence from Star Carr, there was no evidence of fish in the diet. Dried mud inside one of the Aveline's arm bones still contained pollen, indicating a relatively open landscape with birch, pine and grasses.

At this time sea levels had not yet risen enough to re-detach Britain from the Continent or to form the Bristol Channel and, at 80–100 kilometres (50–60 miles) from the coast, Gough's Cave and Aveline's Hole were considerably further inland than they are today. We saw from Star Carr that these early postglacial settlers were impacting and interacting with their environment through the removal and use of timber and the firing of reeds, but this was only on a tiny scale compared with what was to follow. Nevertheless, the people of the Mesolithic began the modification of our landscape that has culminated

Mitochondrial DNA has been recovered from ten Neanderthal fossils, confirming that these ancient humans represented a separate genetic lineage from everyone alive today

in Britain's green and pleasant land. Given the story of interrupted occupation we have seen repeated many times through more than half a million years of human history in Britain, is there at last a genealogical link, however tenuous, between the pioneers who began to enter the peninsula of Britain as the ice finally retreated some 11,500 years ago and the Britons of today?

One line of evidence we can turn to is DNA. In the last chapter, we saw how a special kind of DNA, mitochondrial DNA (mtDNA), has been recovered from ten Neanderthal fossils, confirming that these ancient humans represented a separate genetic lineage from everyone alive today. As we also saw, mtDNA has been used to reconstruct the early evolution and global dispersal of modern humans, and all of us alive today fall into one of the 'haplogroups' descended from the original African mtDNA female ancestor's genetic material. The different haplogroups of Europeans have been identified by the letters H–K and T–X, and the geneticist Bryan Sykes of Oxford University took this a stage further by giving them women's names since, by implication, each of the founders of the haplogroups would have been an Ice Age woman, one of the daughters (many generations removed) of the hypothetical African Eve. Sykes has imaginatively called these Seven Daughters of Eve Helena (she supposedly lived in the Pyrenees), Jasmine (Syria), Katrine (Venice), Tara (who apparently had the good taste to live in Tuscany), Ursula (Greece), Valda (Spain) and finally Xenia (the Caucasus Mountains), although he admits that their precise places of origin are almost as fanciful as the names he has given them.

Using an average mutation rate for mtDNA, it is possible to estimate (with a fair margin of error) the approximate time of origin of the different haplogroups or, as Sykes calls them, clans. U(rsula) would be the most ancient, deriving from some of the first Cro-Magnons, about 45,000 years old, and now present in about 11 per cent of people all over Europe, particularly in the north-west. X(enia) is the second oldest of the seven European clans, dating from about 25,000 years ago, from just before the Last Glacial Maximum, with about 7 per cent of Europeans. H(elena) is the largest and most widespread of the

European clans, dating from about 20,000 years ago and representing about 41 per cent of Europeans, especially common in the Basques of northern Spain and southern France. V(elda) is of similar age but is present in only about 4 per cent of native Europeans, mainly in the north-west, including in the Saami people of Norway and Finland. T(ara) is most common in southern and western Europe, especially the western parts of the British Isles, and represents about 10 per cent of Europeans. K(atrine) also shows a frequency of around 10 per cent, dating from about 15,000 years ago, now most common in the Alps and central and northern Europe. J(asmine) is the second largest of the European clans at about 12 per cent, and the youngest in origin, at about 9,000 years ago. Sykes believes this clan originated in the Middle East among early farmers and spread across Europe with the agricultural revolution of the Neolithic (New Stone Age). He has since added one more, called Ulrike and related to U(rsula). Although originating at about 18,000 years, this strain is present in only about 2 per cent of Europeans, and is found mainly in northern and eastern Europe.

In 1995, Bryan Sykes began working with me in an attempt to recover ancient DNA from the Cheddar fossils, both from the Creswellian layers and from Cheddar Man. Animal bones were tested first and registered that DNA was preserved, so an attempt was made on one of Cheddar Man's foot bones, but without success. In 1997, a different approach was used. The base of the crown of one of his molar teeth was drilled in a 'clean' laboratory, and the resulting tissue extracts were biochemically amplified (copied many times over) to see if human DNA could be identified. The preliminary results received wide publicity through a television series on archaeology, since it transpired that Cheddar Man apparently had DNA of haplogroup U(rsula), and moreover, when local samples were collected in Cheddar village, there were a number of close matches to the sequence, including the mtDNA of the headmaster of the local primary school, Adrian Targett. The results have been discussed widely, but with much misrepresentation or misunderstanding. The one thing we can be pretty confident about is that Targett had not received his mtDNA from Cheddar Man. As we discussed earlier, mtDNA is passed down through females, and thus even if Cheddar Man had fathered children none of them would have inherited his mtDNA. Any link between Cheddar Man and Targett would have to have been through a common female ancestor, at minimum through Cheddar

OPPOSITE: At about 10,000 years old, the skull of Cheddar Man represents one of the oldest real links to the people of Britain today. A hole in his forehead was the site of an infection that may have killed him.

Man's mother (and then a sister), but much more likely through a female ancestor who lived even further back in time, and probably elsewhere – perhaps on now submerged territory such as Doggerland, or even on the mainland of Europe.

And there is one more complication. Since 1997, a great deal more work on ancient DNA has been carried out, and this shows how serious a problem contamination is. Every person who has touched the Cheddar fossils could have left their DNA on them as well, whether in the form of skin flakes or even spittle, as they talked. However carefully the extraction process is carried out, and however clean the lab, the amplification process is so good at locating and copying minute strands of DNA that any contaminants will be copied, as well as any genuinely ancient DNA. If Cheddar Man had DNA as different from recent Europeans as the Neanderthals were, it could certainly be distinguished. But as we have seen, in terms of their mtDNA, recent Europeans have all descended from the same few women, and this means we can no longer be sure that the Cheddar Man result is authentic evidence of a direct link back to his Palaeolithic ancestors (and forward to his living clan member Adrian Targett) rather than the result of contamination by one of the 11 per cent of fellow Europeans who carry the same mtDNA type today. But these doubts should not obscure the very real significance of Cheddar Man to British history. He is not, of course, the oldest human inhabitant of Britain, by hundreds of thousands of years. But as we have seen, the stone tool makers of Pakefield, the butchers of Boxgrove, the Swanscombe woman, all represent people whose hold on British soil was so vulnerable and tenuous that their descendants lost their grip on settlement within a few thousand years. And because they were part of an early dispersal from Africa that did not give rise to modern humans, they have no descendants anywhere in the world today. Even the modern human tribes of Cheddar Man's predecessors in Gough's Cave, a few thousand years earlier, whose butchered bones lay close to his, were apparently swept away when the Younger Dryas brought the ice caps back to Britain. They have no continuity in Britain with the Britons of today, either. It is only when we arrive at the Holocene, the 11,500-year-long interglacial that we live in today, that we can really speak of an unbroken chain of people who connect to their descendants on this island. And Cheddar Man is probably one of those links in the chain, whether his DNA shows it or not. While he and his contemporaries lived in a peninsula of Europe,

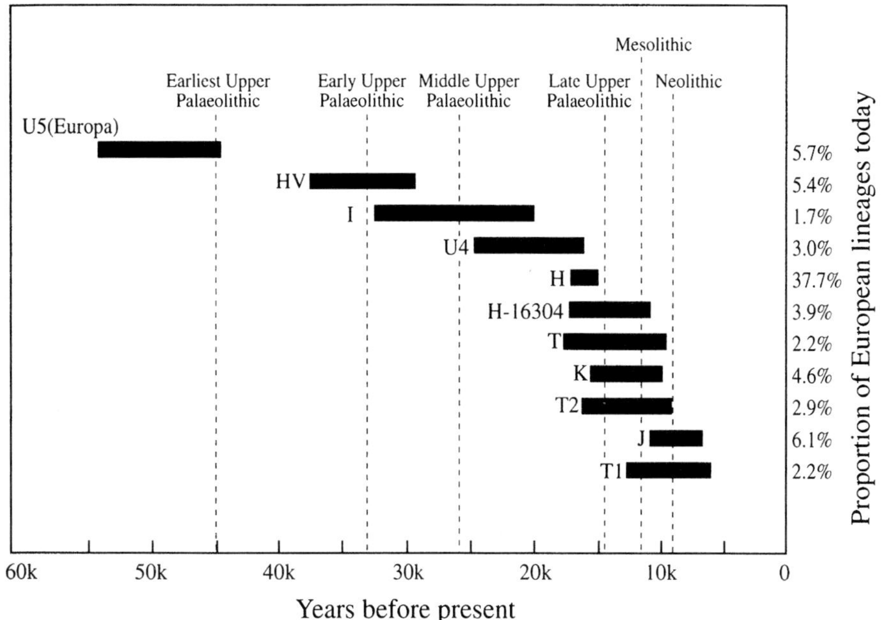

ABOVE: A more complex picture of European mitochondrial DNA history than Eve's seven daughters is shown in this diagram, which estimates the time of appearance of European mtDNA lineages. The variants of U relate to 'Ursula', but additional groups such as 'I' are also shown.

this was not to last, for as the melting of the glacial ice caps accelerated, sea levels rose all over the world. About 9,500 years ago the Irish Sea and the Strait of Dover began to open up, and by about 7,000 years ago, the last vestiges of a land bridge between Britain and France were submerged. The effects of the rise in sea level were compensated for in glaciated regions by the lifting of the weight of the ice caps. These had depressed the northern land masses by hundreds of metres, and the bounce-back following their melting is still occurring today. Since the end of the Ice Age, eastern Canada has risen some 900 metres, Scandinavia 700 metres and eastern Scotland 250 metres: this may explain why skeletons of whales and seals have been found 30 metres above present water levels in the Firth of Forth. Scandinavia has been rising by about 1 centimetre a year, and even shorelines occupied by the Vikings are now some 8 metres above sea level.

Isotope data produced by AHOB member Michael Richards and colleagues show that European Cro-Magnons were using more aquatic and marine resources than Neanderthals did, yet the evidence from Gough's Cave and Star Carr does not seem to show this, beyond the presence of the bones of some water birds. However, another British site at the end of the Ice Age gives us the first evidence of the really intensive use of marine resources. Mike, working with

OPPOSITE: A 10-metre trail of Mesolithic human footprints at Formby Point.

AHOB and other colleagues, conducted an isotope analysis of human bones from Kendrick's Cave near Llandudno in north-west Wales. The Kendrick's people lived at about the same time as the Creswellian inhabitants of Gough's, but at the very limits of the range of people at that time. Like the people at Gough's they hunted horses, but isotopes in their bone collagen suggest that they could have been getting as much as a quarter of their protein from the sea, in the form of fish, shellfish and marine mammals such as seals.

This level of reliance on the sea is exceptional for the Old Stone Age, but five thousand years later, early in our present interglacial, such dependency was widespread across western Europe, from Scotland and Denmark in the north to Portugal in the south. In some sites, massive accumulations of shellfish middens and isotope data suggest that marine foods could have made up almost all of the diet. Some experts see this Mesolithic reliance on a single resource base (the sea) as a prelude to the similar narrowing that followed in the subsequent Neolithic period, with the switch to agricultural and pastoral foods. Evocative traces of the importance of the coasts and estuaries to the Mesolithic people of Britain are preserved in trails of footprints that can be seen today in the estuaries of the Severn and Mersey.

Eight thousand years ago the Severn Estuary did not exist and the river flowed through a forest-covered plain into a bay near what is now the island of Lundy. At Goldcliff, when the Bristol Channel tides are very low, beautifully preserved footprints of cranes, deer, aurochs and humans are briefly exposed. The shoeless human prints belong to adults and, particularly, to children, and there are clearly many exposed footprint surfaces layered one on top of the other, perhaps even representing annual cycles of flooding and drying out of the salt marshes. Pollen, wood and preserved traces of vegetation paint a vivid picture of an environment where the sea was encroaching over vast areas, causing the death of large numbers of trees, the trunks of which survived like upright corpses. Direct evidence of the diet of the people who made the prints comes from butchered bones of wild boar and aurochs, as well as eel bones and burnt raspberry and elderberry seeds, indicating in that case autumn occupation. Even human intestinal parasites have been recovered from these layers, suggesting defecation. There are also charcoal, burnt reeds and larger pieces of burnt wood, showing that where there were surviving forests, they

and surrounding reed beds were probably being cleared by burning (as at Star Carr). The footprints fade out from the time the sea level stabilized about 6,700 years ago, with salt marsh giving way to reed swamp, and then the return of trees in a fen woodland. A slightly later series of hundreds of prints has been found near Formby Point at the mouth of the present Mersey Estuary, with a wider range of animals represented, including once again crane, aurochs and deer, but also unshod horse, dog or wolf, wild boar, sheep or goat, and wading birds such as rail and oystercatcher. The quality and length of the human trails are so good that long toenails are indicated on some prints, and estimates of height and walking speed can be made on others. Some individuals (men?) were about 1.65 metres (5ft 5in) tall, and somewhat smaller individuals (women?) 1.45 metres (4ft 9in), while again many children were represented. Patterns of the prints suggest that women and children may have been collecting razor-shells and shrimps, while males were travelling faster near red deer and roe deer tracks. Whether this was hunting, trapping, or the beginning of animal management is unclear.

But the evidence from several of these Mesolithic sites of larger and perhaps more settled aggregations of people, a greater concentration on particular animal or marine resources, and the beginning of landscape modification in the form of clearing and burning all herald changes which would become even more pronounced in the succeeding and final stage of the Stone Age in Britain – the Neolithic or New Stone Age. In the following and last chapter, we will look at how humans have increasingly impacted and manipulated their environments, and what this means for the future of *Homo britannicus*.

OPPOSITE: A beautifully preserved footprint at Formby, perhaps that of a teenager or woman.

OVERLEAF: A maze of footprints at Formby, including those of wolf or dog, aurochs and human children.

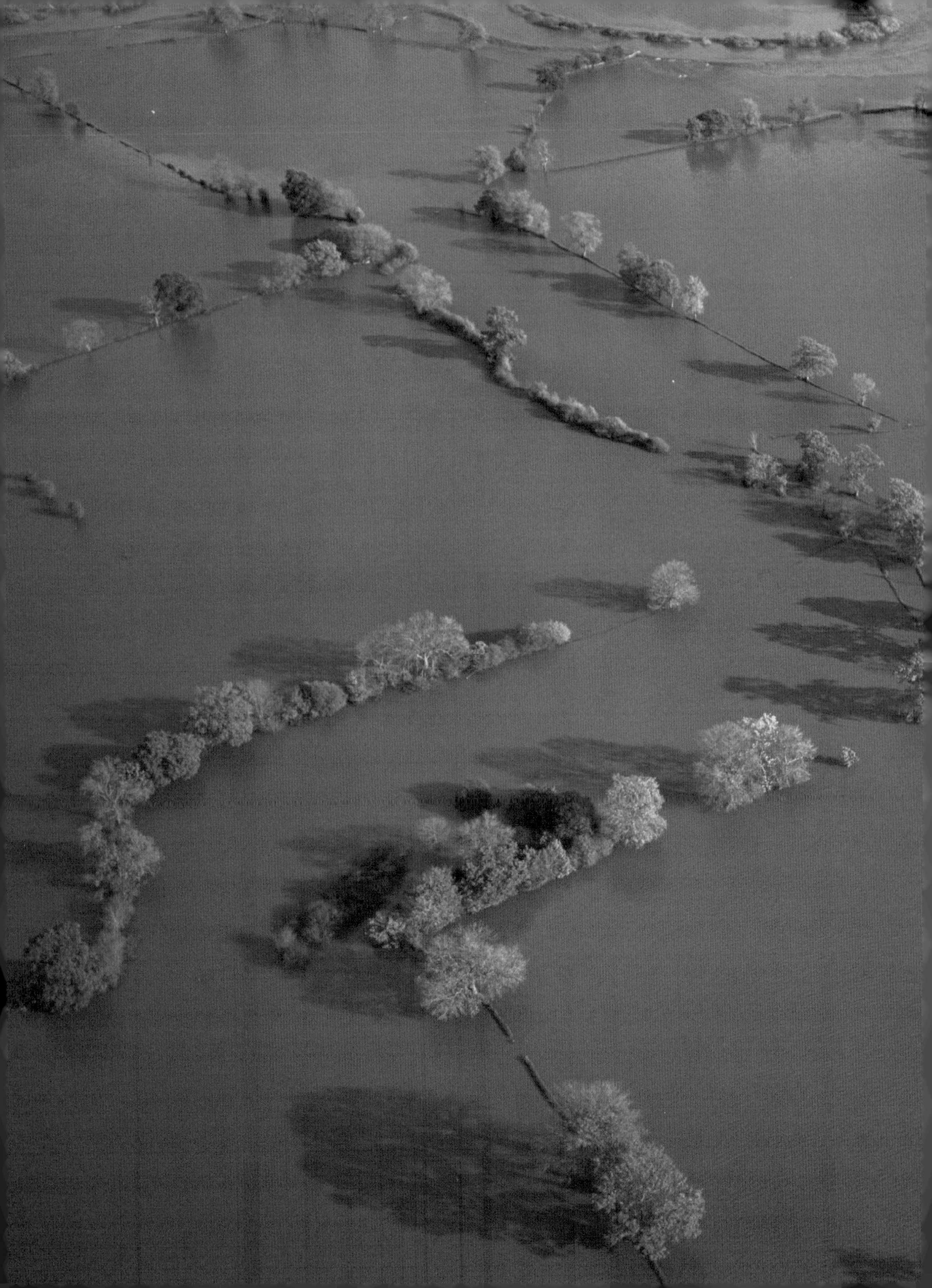

CHAPTER SEVEN
OUR CHALLENGING CLIMATES

PRECEDING PAGES: Flooded fields in Cheshire, UK. The Foresight Future Flooding report published by the Department of Trade and Industry predicts that annual damage from flooding in the UK could rise from £1 billion to £25 billion during the present century.

The early occupation of Britain was often a precarious business, as we have seen. The climate of this green and pleasant land has suffered many slings and arrows, such as the major glaciations of 450,000 and 20,000 years ago that stripped the land bare and covered much of it with an ice sheet up to a mile thick. But long before that, at least 700,000 years ago, people had arrived in a climate warm enough to have subsequently been nicknamed the Costa del Cromer, when pond tortoises, water chestnuts and hippos thrived here. As glaciations grew in their severity, Britain was repeatedly given up to the ice, with at least seven abandonments of settlement by pre-Neanderthals, Neanderthals and, finally, modern humans. In between, there were benign interglacials when people were able to establish themselves, and one 125,000 years ago when, despite the inviting conditions, the English Channel blocked any attempts at colonization.

We have been fortunate enough to live in one of the most stable periods of the last 500,000 years in terms of world climates, and this has allowed humans to spread and settle throughout most of the world, apart from the polar regions, and to enjoy an astonishing growth, from the few million people estimated for the end of the last ice age to our present figure of over 6 billion.

If we look even further back in time, the Earth has suffered yet greater climatic insults. Our planet is over 4.5 billion years old, and after its molten surface had cooled, life appeared more than 3 billion years ago. During the following billion years, the world experienced its first ice age. Although the youthful sun was colder then, Earth's atmosphere was initially high in greenhouse gases such as methane, which kept the surface and its primeval oceans warm. (A greenhouse effect is produced when some of the sun's energy that would otherwise be reflected back from the surface into space is instead absorbed by gases like methane and carbon dioxide and continues to warm the planet.) However, by 2.3 billion years ago, plants making their food through the process called photosynthesis produced a gradual increase in oxygen at the expense of other gases, and temperatures plunged as the greenhouse effect weakened. Life clung on and the Earth recovered until about 750 million years ago when it seems that several even more severe cold events happened, and created a 'snowball Earth'. The oceans froze and the planet's surface was almost entirely covered with ice. While life must have survived in the seas under the ice, it was probably only the effect of volcanoes protruding through the ice and continuing to pour out enough of the greenhouse gases carbon dioxide and methane that broke the pattern. Otherwise Earth might never have emerged from its snowball, and would today look much like Europa, a moon of the planet Jupiter, whose briny oceans are covered by ice miles thick.

The reasons for these astonishing ancient fluctuations are still unclear, partly because so much of the evidence for them is either buried deep or eroded away. The temperature of the young sun may have been less stable, there may have been much more interplanetary dust and debris to interrupt its warm rays, or the tilt of the Earth's axis may have been different from today, making the Milankovitch effect (see pp. 81–2) more extreme. Certainly the composition of the Earth's atmosphere must have regularly changed with the constant interplay between cycles of volcanic activity, unstable weather patterns and the growing

impact of life, while the shifting protocontinents constantly changed the geography of the Earth and its ocean circulations, with land masses sometimes at the poles, and sometimes straddling the equator.

At the transition between the Permian and Triassic periods, about 250 million years ago, the largest known mass extinction occurred, with some 90 per cent of species dying out. Many scientists believe that this period had the warmest environment that life has ever experienced, and that this hot-house world was caused by a sudden release of previously frozen methane hydrates from the cold ocean depths. This huge injection of a powerful greenhouse gas imposed a 5°C rise on an already warming Earth, and in less than 100,000 years, most life had perished. During the Paleoeocene and Eocene, 60-45 million years ago, the continents were beginning to approach their modern configurations, although sea level was some 70 metres higher, and Earth was predominantly warm and humid, with alligators living near the North Pole, palm trees growing in Patagonia, and regions like Europe, North America and Australia covered with subtropical forests. Since then the Earth's climate has deteriorated inexorably, with a mean fall of some 5°C and twice that amount in the Atlantic, and a build up of ice, first at the South Pole and later at the North. Various factors have been implicated in the cooling, two of the most important being a large land mass stuck at the South Pole and, as India crashed into Asia through continental drift, the inexorable rise of the Tibetan Plateau, which changed atmospheric circulation patterns.

In the last million years, the Milankovitch cycles seem to have pushed Earth to ever-greater swings of climate, with the added superimposition of short-term fluctuations lasting a few millennia. The last of these really severe global climatic blips was the Younger Dryas of about 13,000 years ago. This briefly drove people out of Britain one last time, but also set the scene for the modern world by changing environments at a critical point in human history, catalysing the agricultural revolutions in south-western and eastern Asia with the domestication of cereals, animals and rice.

After the upheavals of the Younger Dryas, the Earth's climate settled into the relative stability of our present interglacial, the Holocene. In the more recent past, from about AD 800 to 1300 Europe experienced conditions generally slightly warmer than today, and this led Erik the Red to found colonies on the

OPPOSITE: Europa, a moon of Jupiter, has a briny ocean under its frozen surface. About 750 million years ago, during severe cooling, the Earth may have looked like this.

coast of Greenland, where Vikings fattened their cows in the summer and roamed to hunt seals and reindeer in the winter. Following their French kin, the British grew grapes and developed a thriving wine trade to rival that of their southerly neighbours. But in the 1400s, average temperatures dropped some 2°C in many regions and this cold snap, known as the Little Ice Age, lasted about 500 years, give or take some short ameliorations. The Norse settlements in Greenland had vanished completely by 1500, while British vines soon withered and perished along with hundreds of thousands who starved to death across Europe as their crops failed. At times shipping could not move in the winter when the Thames and the canals of Holland froze over, as depicted in contemporary paintings. In many ways the Little Ice Age shaped Europe irreversibly, influencing social, political and agricultural changes that are still with us today. But during the 1800s, Europe started to warm again, heralding our present climatic era.

ABOVE: Erik the Red (*c.* 950–1003) was born in Norway but later accompanied his father to Iceland. In 980 Erik was forced into exile after killing a man, and his party of explorers made the first known European landfall and settlement in Greenland.

Given the fact that the Earth has experienced such extremes of climatic fluctuations in its history, are the changing climates of today simply part of the natural cycle, or are we interfering with that cycle, for better or worse? Looking at climate changes during the last 200,000 years, and the ongoing Milankovitch cycles, Earth should have been entering a cooling phase now, with the expectation of major glaciation within the next 50,000 years. But many scientists believe that the effect of increasing greenhouse gases in the atmosphere will override any cooling trends in the immediate future. However, that news is not as good as it might seem, because there is a profound danger we will permanently force the Earth's climate out of kilter.

There are those who still doubt the reality of global warming and our role in it, yet the vast majority of scientific data are pointing in the same ominous direction, even if our incomplete understanding of all the parameters prevents precise predictions of how quickly things will change and how bad it will get. A thousand papers on climate issues published in the major science journals between 1993 and 2005 have agreed with the consensus position that global

Arctic sea ice dropped to its smallest ever extent, the Atlantic suffered a record hurricane season, and an unprecedented drought reduced the flow of the Amazon to its lowest known level

warming really is happening, and the seriousness is such that in 2006 the Chief Government Scientist of Britain, Sir David King, argued that global warming is now the single greatest threat facing humankind and our planet. He virtually accepted the inevitability of a 3°C rise, which could put a billion people at risk of starvation because of lost arable land and water shortages, as well as destroying large tracts of tropical forest and half the world's wildlife reserves and corals. Temperatures could increase by a further 1.5° as a result of positive feedbacks in the climate resulting from the melting of sea ice, thawing permafrost and the acidification of the oceans.

So what are the data leading to these worrying conclusions? According to weather records that go back 150 years, nineteen of the twenty warmest years have occurred since 1980, while 1998 was the warmest year since world temperature records were kept and 2003 provided Britain with a new record temperature of 38.5°C. The exceptionally hot, dry summer of 2003 is estimated to have caused about 35,000 additional deaths in western Europe, and analyses suggest that global warming related to elevated carbon dioxide levels has made such summers four times more likely than in a world where the climate is stable. And it is not just Britain and Europe that have been affected, since Canada and Australia also had their hottest weather on record in 2005. Arctic sea ice dropped to its smallest ever extent, the Atlantic suffered a record hurricane season, and an unprecedented drought reduced the flow of the Amazon to its lowest known level. These developments are worrying enough, but there is evidence that we have not yet felt the full force of global warming because of a phenomenon called global dimming. Against expectations, and in seeming contrast to rising temperatures, the measured amount of sunlight reaching the Earth's surface has fallen over the last fifty years. Research suggests that this is because of the high levels of polluting particles in the atmosphere from burning aviation fuel, coal, oil and wood. In clouds, these particles seed water droplets, increasing their reflectance and reducing the penetration of sunlight. Thus the warming effect

OVERLEAF: Abraham Hondius painted *A Frost Fair on the Thames at Temple Stairs* in about 1684.

ABOVE: These reconstructions from satellite images show changes in the extent of sea ice in the Arctic: (*left*) 1979; (*right*) 2003.

of the increase in polluting greenhouse gases has so far been partly counteracted by the parallel increase in particles. As we reduce particle pollution, as we must, the brakes will be off for the full force of the greenhouse gases.

And the proportion of greenhouse gases in the atmosphere, especially carbon dioxide, has risen relentlessly – CO_2 levels are now over a third greater than before the Industrial Revolution and higher than any level recorded in gas bubbles trapped in Antarctic ice over the last 650,000 years. Already we are at a degree of global warmth approaching the maximum of any interglacial in the last 500,000 years, but if current trends in greenhouse gas emissions continue unabated CO_2 is projected to be double that of pre-industrial levels by the end of this century, when global temperatures are likely to rise by about 3°C. Projections suggest that before then Britain will return to the level of Mediterranean warmth found at Pakefield 700,000 years ago. That may sound appealing. In the south British farmers could grow oranges, lemons, avocado and melon, our vineyards could rival those of France, and our beaches those of Spain. The landscapes of Britain would change dramatically as familiar forests and woodlands of oak and beech give way to Spanish chestnut, Turkish hazel, and olive trees, along with the plants and insects that accompany them. This may not seem so bad, but northern Britain could suffer increased rainfall, and warmer temperatures could lead to the spread of exotic diseases such as malaria, yellow fever and West Nile fever, as well as increased deaths from heat, skin cancer and food poisoning. Moreover, as the environmentalist James Lovelock put it, 'our nation is now so urbanized as to be like a large city and we have only a small acreage of agriculture and forestry. We are dependent on the trading world for sustenance; climate change will deny us regular supplies of food and fuel from overseas.' Worse still, after 2100 the climatic clock could start to be wound back 60 million more years to the subtropical levels of the Eocene, and beyond that, if the process cannot be stopped, a runaway greenhouse effect could return us to the scorching heat of the hot-house Earth of 250 million years ago, when most life on the planet was eliminated.

Future climate change is not just about rising temperatures. Altering one critical parameter retunes the whole system in complex ways so that atmospheric circulation, ocean systems, wind, rain and snow patterns will all be altered. The thawing of permafrost in the north shows an example of the possible knock-on effects. Permafrost is ground that is frozen throughout the year, and it can run up to a mile deep in places like Siberia and Alaska – it makes up about a quarter of land area in the northern hemisphere. Already, land surfaces in parts of Alaska have become unstable through melting permafrost, causing roads and houses to subside into newly created morasses of mud and marshes, and producing 'drunken forests', where the trees lean at wild angles. Some permafrost is hundreds of thousands of years old, and it is estimated that the preservation and decay of past vegetation has accumulated about 450 billion metric tonnes of carbon in the Earth's permafrosts – nearly a third of the total stored in soils globally. Should that carbon start to be liberated by melting and subsequent bacterial action, even larger quantities of CO_2 and the much more powerful greenhouse gas methane would be added to the already warming atmosphere. The northern oceans would be seriously affected too, as melting ice and accelerated drainage would pour increasing quantities of fresh water into them; such runoff has already increased by some 7 per cent since the 1930s.

The Arctic itself is heading towards a seasonally ice-free state for the first time in over a million years and the duration of seasonal snow cover is reducing noticeably, both of which reinforce global warming through the loss of an effect called albedo, the reflection back of solar radiation by snow and ice, keeping ground temperatures low in a feedback effect. This cycle is being disrupted as the degree of reflectance decreases, since newly exposed soils, vegetation and sea are darker than snow and ice, thus they absorb, store and then release the warmth. And a northward spread of trees and shrubs may itself add a seasonal rise of at least one degree to Arctic spring temperatures. The prospect of a

radically different climate is a direct threat to the plants, animals and humans native to the Arctic. For example, polar bears live on ice throughout the year, making it their base to hunt, breed and raise their young. As the ice packs shrink and break up in the summer thaws, bears swim between ice floes to continue hunting. However, the Arctic ice caps have recently receded about 200 miles further north during thaws, forcing the bears to undertake far longer swims between floes of as much as sixty miles, swims through waters that may be stormier or with stronger currents. For the first time, significant numbers of adult carcasses are being found in the water in regions such as northern Alaska, bears that have either drowned or died of exhaustion; and the number of bears is declining right across their range. The Inuit, who are indigenous to the north American Arctic, have now petitioned the Inter-American Commission on Human Rights, accusing the US of destroying their traditional way of life and threatening their future through fuelling climate change, leading to the total degradation of their environment.

ABOVE: Polar bears are threatened with extinction as the ice floes on which they live and hunt for much of the year are breaking up under the influence of global warming.

OPPOSITE: 'Drunken forests' are being created in many northern regions as permafrost melts.

Ironically, though, the place where global warming is showing its effects most strongly is at the other end of the Earth, the virtually uninhabited Antarctic. In the past forty years records show that this continent has warmed faster than any other. Sea ice has shrunk by 20 per cent since 1950, the west Antarctic ice sheet is dissolving at an accelerating rate, glaciers are thinning and retreating, and the Ross Sea is freshening through increased melting. In contrast, however, greater evaporation is leading to thicker snows in east Antarctica.

We have seen how the past ebb and flow of the ice caused direct effects on sea level, with high stands in interglacials and falls of over 100 metres compared with present levels during the coldest stages. In addition, there were the effects of the growth or decay of heavy ice caps on the altitude of the land itself. If the most vulnerable ice sheets of today – those in Greenland and west Antarctica – start to melt with rising temperatures, this will herald a devastating increase in sea level. Many of the world's largest cities such as London, New York, Kolkata,

OPPOSITE: The Thames Barrier at Woolwich Reach has successfully protected central London from serious flooding since 1982. However, it may cease to be effective by about 2030.

Cape Town, Hong Kong and Tokyo have been built on coasts, river estuaries or small islands. London was last seriously flooded in 1928, and the Thames Barrier began operation in 1982, built to cope with the risks known at that time. This barrier is a series of steel gates between concrete towers that can rapidly be closed if London is threatened by a tidal surge. It has protected the city very effectively for the past two decades, but it is now being raised with increasing frequency due to elevated sea levels, storms and tidal amplitude. It is likely to be ineffective after 2030, and to protect London from a possible £30 billion financial damage, not to mention the social, political and human costs, a new 10-mile-wide barrier across the Thames estuary between Essex and Kent is now under discussion.

On the other side of the North Sea, the Dutch are adopting a very different approach. They have worked ceaselessly for centuries to protect their lands from flooding (there are over 10,000 miles of dykes, barriers and drainage channels), but the devastating floods of 1995 finally showed the futility of this constant battle, and forced what can really be called a sea change. Over the next half-century, the Dutch will allow an area more than twice the size of Greater London to flood as part of a strategy to live with and adapt to the encroaching waters, and an increasing number of amphibious houses, shops and even greenhouses are being built, which can function equally well on the land or (anchored by metal posts), floating on water.

Part of the present modest sea level rise is actually due to the thermal expansion of the oceans because of warming, but a predicted rise of 3°C would not only see further expansion, but probably also the beginning of the meltdown of the massive Greenland ice cap. Even a rise of one metre from a partial melt would threaten not only London, but Hull, Liverpool, the south coasts of England and Wales, and East Anglia, as well as the Netherlands, many cities on the US eastern seaboard, and a host of islands. For example, the Maldives in the Indian Ocean have 360,000 inhabitants living on specks of land that mostly lie less than a metre above sea level.

If the whole ice cap disappeared, there would be an even more serious sea level rise of some seven metres, enough to engulf coastal and low-lying lands occupied by at least two billion people, including the deltas of the Nile, Niger, Ganges, Mekong, Amazon and Mississippi. The devastation of New Orleans by

E
D
C
7
8

Hurricane Katrina in 2005 was predicted in detail long before it happened, and it was only the speed and scale of the evacuation programme that saved many thousands of lives. The disaster occurred, of course, without greatly elevated sea levels, but it may still be attributed to the effects of global warming on the Earth's weather systems. Although the number of hurricanes forming in the mid-Atlantic during the last few years has been high, it is unclear whether this in itself is related to higher temperatures rather than natural variation. But the growing intensity of the hurricanes in terms of power and associated deluges fits perfectly with models that show how higher temperatures feed more energy into the weather systems that create and fuel the hurricanes. Thus as greenhouse gases build up in the atmosphere, the average strength of hurricanes should increase, and this will be part of a more general trend for our weather to become more extreme in terms of the levels of wind, rain, snow and, in the opposite direction, drought.

OPPOSITE: 2005 was a record year for tropical storms such as this hurricane, viewed from space. Global warming will put increasing amounts of energy into the Earth's climate system, leading to more extreme weather around the world.

So it seems that we are destined to live in a warming world – or are we? If the increase of melt water from thinning ice cover and thawing permafrost leads to a build-up of cold freshwater in the North Atlantic, we could instead suffer a Heinrich Event (named after the German scientist who first recognized them) lasting a few centuries, when the North Atlantic suddenly chills, and there is a massive southerly flow of icebergs (an event over-dramatically depicted in the film *The Day after Tomorrow*). As we saw in Chapter 5, the Atlantic Heat Conveyor that warms north-western Europe by carrying Gulf Stream waters on its surface has periodically shut down, with devastating effects on the climate of Britain and surrounding regions. Air temperatures over the Atlantic could sink by as much as 10°C and in western Europe by 4°C. Careful research is being conducted to see if there has been any weakening of the Atlantic Heat Conveyor, and the worrying news is that such an effect has already been detected. A recent study of a transect across the subtropical Atlantic using data from the last fifty years suggests that the Conveyor there has weakened by more than 30 per cent.

While more data from different depths and locations are now anxiously awaited to confirm the pattern or not, the situation does not bode well, and might give Britain a climate that it has never experienced before, one that is extremely continental (that is, one that matches the climates found deep within continents, without the moderating effect of the oceans). Continuing global

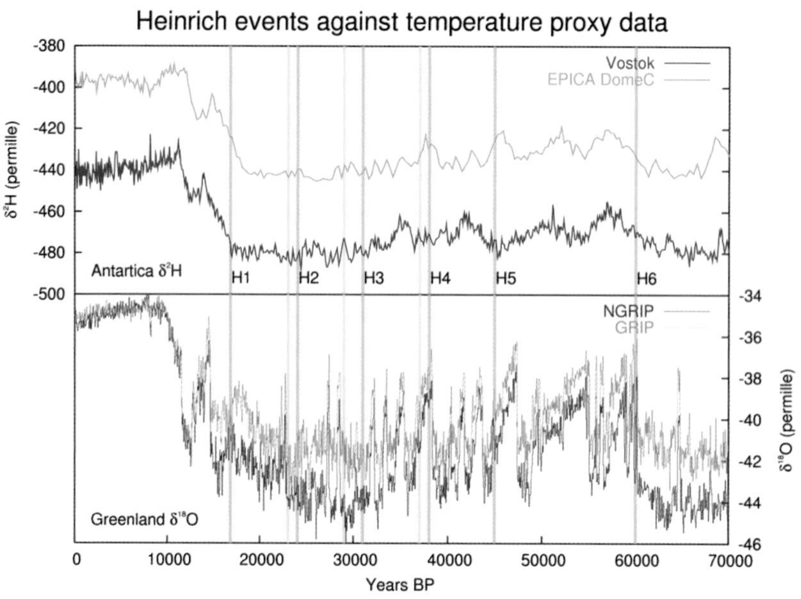

ABOVE: The Atlantic has repeatedly been chilled by 'Heinrich Events', where armadas of icebergs have suddenly flowed southwards, at times even reaching the mouth of the Mediterranean.

warming of the atmosphere and land would mean our summer temperatures might continue to climb, first to Ipswichian or Hoxnian peak levels (about a degree higher than today) then to those of Pakefield (about 2°C higher) and beyond. However, in the winter, the effect of the icy Atlantic would dominate, giving us a highly seasonal climate more like that suffered by the Neanderthals at Lynford, or like Labrador today. However, based on past performance, Heinrich Events have generally been very short, so this strange situation might last only for a few centuries before the Gulf Stream switched back on, stoking up Atlantic temperatures again. As we saw in Chapter 5, some of these switches of Atlantic circulation have happened in the blinking of an eye, geologically speaking, so if the Atlantic observations presage the real thing, we may only have a few more years to prepare ourselves.

So what should we do in the face of these enormous threats to our future? In his *Four Quartets* T. S. Eliot says, 'human kind cannot bear very much reality', but we must avoid the tendency to bury our heads in the sand, or deny the truth of what we have been doing to our planet. As far back as 1989, a prominent world leader addressed the UN with these words:

While the conventional, political dangers – the threat of global annihilation, the fact of regional war – appear to be receding, we have all recently become aware of another insidious danger. It is as menacing in its way as those more accustomed perils with which international diplomacy has concerned itself for centuries. It is the prospect of irretrievable damage to the atmosphere, to the oceans, to Earth itself.

What we are now doing to the world, by degrading the land surfaces, by polluting the waters and by adding greenhouse gases to the air at an unprecedented rate – all this is new in the experience of the Earth. It is Mankind and his activities that are changing the environment of our planet in damaging and dangerous ways.

The result is that change in future is likely to be more fundamental and more widespread than anything we have known hitherto. Change to the sea around us, change to the atmosphere above, leading in turn to change in the world's climate, which could alter the way we live in the most fundamental way of all. That prospect is a new factor in human affairs. It is comparable in its implications to the discovery of how to split the atom. Indeed, its results could be even more far-reaching.

The evidence is there. The damage is being done. What do we, the international community, do about it? The environmental challenge that confronts the whole world demands an equivalent response from the whole world. Every country will be affected and no one can opt out. Those countries that are industrialized must contribute more to help those who are not.

The work ahead will be long and exacting. We should embark on it hopeful of success, not fearful of failure. Darwin's voyages were among the high-points of scientific discovery. They were undertaken at a time when men and women felt growing confidence that we could not only understand the natural world but we could master it, too. Today, we have learned rather more humility and respect for the balance of nature. But another of the beliefs of Darwin's era should help to see us through – the belief in reason and the scientific method.

Reason is humanity's special gift. It allows us to understand the structure of the nucleus. It enables us to explore the heavens. It helps us to conquer disease. Now we must use our reason to find a way in which we can live with nature, and not dominate nature. We need our reason to teach us today that we are not – that we must not try to be – the lords of all we survey. We are not the lords, we are the Lord's creatures, the trustees of this planet, charged today with preserving life itself – preserving life with all its mystery and all its wonder. May we all be equal to that task.

OPPOSITE: The Aletsch Glacier in Switzerland. The impact of global warming is being felt right across the European Alps, and future scenarios predict even greater changes.

Those were the prescient words of Margaret Thatcher, but not only were they unheeded by most of the world leaders who heard them, they were sadly also ignored by the Prime Minister herself, whose government proceeded to block or water down proposals to deal with the problems, as well as embark on the most massive programme of road-building the country had ever seen, implicitly encouraging pollution by motor vehicles. Although she would not bite the bullet, Mrs Thatcher at least acknowledged – seventeen years back – the stark reality that we have been conducting an uncontrolled experiment with the Earth's climate system for at least three hundred years. In fact, from examining ice cores that record atmospheric changes some scientists believe the rot set in as far back as 8,000 years ago, when farming really took off and human numbers started to grow significantly for the first time. Before then, humans lived as hunter-gatherers, and were more or less obliged to follow the natural rules that limit the population densities of any animals at the top of their food chains. But it's argued that once people in Europe and Asia began clearing forests for crops and pastures, they started to reverse a natural decline in CO_2 content, and by 5,000 years ago had done the same with methane levels through increasing herds of livestock and by flooding fields for rice culture. In which case these early farmers may have inadvertently warded off the cooling that tends to happen as an interglacial passes its peak, as our present one might otherwise have done. If that is so, we have been fortunate, but our luck is quickly running out as we move into completely unknown territory, tinkering with the Earth's climate machine.

The effects of global warming will indeed be global, which is why no peoples or nations can afford to be complacent about it. Europe is currently devouring the world's natural resources and producing pollution at twice the global average, although still well below US levels. Research suggests that its mountains and southern lands will be the hardest hit by climate change. Ten per cent of Alpine glaciers disappeared during the summer of 2003 alone, and at current rates 75 per cent of Switzerland's glaciers will have melted by 2050.

Snow lines are getting higher and less snow is being stored through the year, which will seriously impact hydroelectric power stations, and more obviously the skiing industry. Some Mediterranean cities are already intolerably hot and polluted in the summer, but this will get worse, and there will be an increase in forest fires and water shortages. Farmers in the north would benefit, at least for a time, from exploiting crops now grown in the south, but they are also likely to suffer much less predictable weather and a greater risk of flooding. With increased summer heat, rationed water supplies and arid lands, the Mediterranean region may be increasingly abandoned as millions of people move north, creating enormous new social and political pressures.

The United States is far and away the biggest contributor to greenhouse gas emissions, and perhaps with the New Orleans floods it is starting to reap what it has sown, since warmer seas fuel more powerful hurricanes. Cities will increasingly suffer from water shortages and baking summers, resorts which rely on skiing could see snows retreat by a thousand feet, and tropical diseases like dengue fever will spread in the south as higher temperatures and new wetlands create ideal breeding conditions for disease-carrying insects. Although increased rainfall in some areas will benefit agriculture, life for city-dwelling Americans will get increasingly uncomfortable as floods, heat and drought add to the pressures of crime, traffic congestion, compromised air and water quality, and decaying infrastructure. Stronger hurricanes will batter the south-east, downpours, ice and snow melts will feed flash floods, and an eventual sea level rise of seven metres would see Florida turned into an island. Commenting on President George W. Bush's denial of the reality of global warming, Anthony Janetos of the World Resources Institute said, 'Because the markets don't have an efficient way to value what the seas, the rivers, the mountains and the forests actually contribute, they effectively ignore them. But you do that at your peril.'

China is likely to become the world's leading economic force, if damage caused by climate change doesn't get in the way, but there is a great danger that it will repeat most of the mistakes that western industrialized nations have made in contributing to environmental degradation and greenhouse gas emissions. Chinese experts have warned that since 1950 there has been a gradual reduction in precipitation nationwide, since 1960 the volume of the six largest rivers has steadily declined, and since the 1980s its northern provinces have regularly suffered droughts and flash floods. Moreover they predict that two thirds of China's high-altitude ice fields (about 15 per cent of the Earth's total) would melt by 2050 if current trends continue. Already there are noticeable changes in vegetation patterns, and if climate change drastically affects rice production, one sixth of the world's population would suffer.

Our evolutionary homeland of Africa may be one of the most severely hit by the effects of global warming. The northern half of the continent is likely to suffer even worse droughts in the twenty-first century, with perhaps a further fall of 30 per cent in rainfall compared with last century, disastrous when about 70 per cent of the population depend directly on rain-fed agriculture. The south may not fare as badly, but as in other regions, weather throughout Africa is likely to get much more extreme. One of the icons of Africa, Tanzania's ice-capped Mount Kilimanjaro, is rapidly losing its glaciers, and the country's climate researchers have noted another growing trend: areas that usually get two

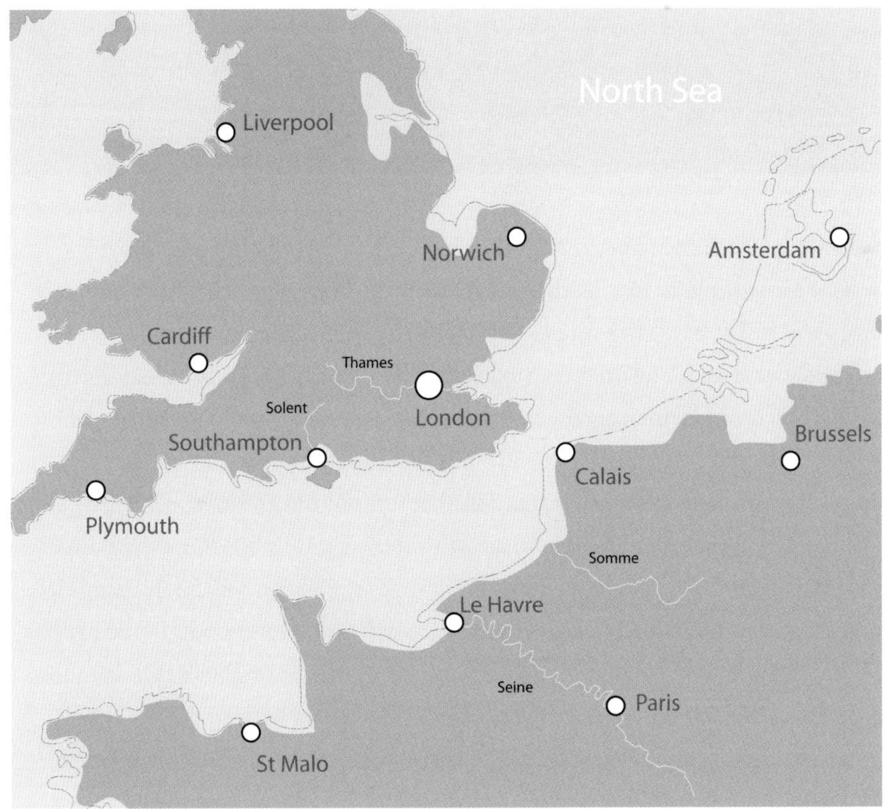

LEFT: If the Greenland Ice Cap begins to melt, sea level will rise by up to 7 metres, as shown in this projection. Many low-lying regions of Britain will be inundated, while most of the Netherlands would be under water.

ABOVE: Ice on the summit of Mount Kilimanjaro, Africa's tallest mountain, is rapidly disappearing.

rainfalls in the year are getting more, while those that get only one rainy season are getting far less, leading to growing aridity. Maize, the main staple crop, will be hit hard, and forests, rivers and hydroelectric power will be seriously depleted. The accelerating clearance of ancient forests in regions like the Congo basin is not only destroying rich biodiversity but is altering local climate, rainfall patterns, rivers and soils, fuelling the spread of deserts.

The problems of Africa remind us that it is not only humans who are being affected by climate change. Scientists estimate that global warming may drive more than a million species to extinction by 2050, and this will be the *coup de grâce* for what has already been termed the Sixth Great Extinction in the history of life, and the only one that can be attributed to the action of a single species. The end of the last Ice Age saw the species of the world hit by a double blow: massive climate and environmental changes in a very short time as the world emerged from an unstable end to the last glaciation, plus the growing impact of humans, first as increasingly efficient predators and then as farmers, with associated environmental change and human population growth. The large and diverse mammal megafauna of the Americas and Australia, evolved over millions of years, went under in a few thousand, and Europe entered an interglacial without species like elephants, lions and rhinos for the first time since these creatures began their Pleistocene occupation of the continent. These disruptions to the natural world have continued right through to the present day, with the prospect of a third blow in the form of very rapid climate change, in a new direction and on an even greater scale. Many species that manage to survive will do so only by major changes to their ranges or life cycles. This is already showing itself in the oceans, where fishing stocks in the Atlantic have moved north: cod and haddock are now found north of Iceland instead of to the south, while warmer-water species such as monkfish are appearing there for the first time. In Britain, spring is starting about two weeks earlier than it did fifty years ago, with consequences for the first appearances of many species: Red Admiral butterflies and bumble bees have been sighted in January, and grass is now growing throughout the year. Some of our colder fauna and flora such as capercaillie and snow bunting, and arctic alpine flowers, are likely to disappear completely, with new immigrant species from the south taking their place. But globally some species have narrower tolerances, which is why many corals are dying as seas get too warm for them, or the chemistry of the oceans undergoes

LEFT: Birds adapted to colder conditions such as the capercaillie (*left*) and snow bunting (*right*) are likely to disappear from Britain as global warming takes effect.

change. And if microscopic plankton is also adversely affected, as some experts fear, not only will this lead to extinctions through many food chains, but it will interfere with one of the main natural carbon storage systems, as I explain later.

So what can we do in the face of this daunting catalogue of past, present and future woes? And are there any lessons from the past that might help us deal with what may be to come? It is possible that in the medium to long term we will be able to develop efficient and stable carbon sequestration – that is, underground or underwater storage of carbon direct from power station emissions, or extracted from the atmosphere to reduce the level of greenhouse gases. Possibilities include pumping CO_2 deep underground, or converting it into mineral carbonates. Biological storage might also help – by adding iron sulphate to the oceans, the growth of plankton, including algae, would be encouraged (but with the danger of altering ocean chemistry). The former store carbon in their bodies, and these sink naturally to the seabed when they die, while the latter reduce CO_2 levels and increase oxygen levels through photosynthesis. Other medium- to long-term solutions would be the successful development of virtually emission-free nuclear fusion power, by duplicating on Earth the process by which the sun creates energy, and finding ways to turn down the Earth's thermostat by increasing the reflectance of the Earth's clouds, so that less of the sun's energy penetrates them to add to global warming.

But these are all risky propositions to rely on, for one reason or another, and most scientists and environmentalists (plus a growing number of politicians) agree that we must act immediately to reduce the emission of greenhouse gases, and ideally return atmospheric levels at least to those of the last century. No less a figure than broadcaster and naturalist Sir David Attenborough recently confessed, 'I'm no longer sceptical. I think climate change is the major challenge facing the world.' A powerful reason behind his speaking out was the thought that his grandchildren would ask him why he had known about global warming and yet did nothing. Everyone has a responsibility to reduce their own 'carbon footprint': using less energy, more efficient and ideally renewable energy (such as solar, wind or tidal), and leading less polluting life-styles. This not only means driving the car less, or not at all, but also cutting back on air travel when it has never been so cheap (but never so costly for our planet). While this will come as a shock to an industrialized world that seems to expect permanent growth in energy consumption, wealth and mobility, we all need to make changes now, or generations to come will pay the price for our selfishness. And since every human is, unavoidably, a consumer and a polluter, human population growth itself must be curtailed and eventually reversed, before nature performs the task for us. The challenges may seem daunting, but if everyone contributes now by making small and consistent changes to their life-style, we can make real and measurable progress towards a better future for our species and the world. And we have a major responsibility in ensuring that our governments put the problem of global warming at the top of their list of priorities, not only when they are trying to get elected, but throughout their terms of office. We also need informed debate on the critical question of building a new generation of nuclear reactors that will certainly reduce the consumption of other sources of energy and carbon emissions, but at the cost of accumulating radioactive waste instead. Unless we address the threat of global warming properly through other measures, it may be forced on us.

James Lovelock takes an especially grim view. He states:

> We are responsible and will suffer the consequences: as the century progresses, the temperature will rise 8°C in temperate regions and 5° in the tropics. Much of the tropical land mass will become scrub and desert, and will no longer serve for regulation; this adds to the 40 per cent of the Earth's surface we have depleted to feed ourselves. Before this century is over billions of us will die and the few breeding pairs of people that survive will be in the Arctic where the climate remains tolerable. So what should we do? First, we have to keep in mind the awesome pace of change and realize how little time is left to act; and then each community and nation must find the best use of the resources they have to sustain civilization for as long as they can. Civilization is energy-intensive and we cannot turn it off without crashing, so we need the security of a powered descent. We will do our best to survive, but sadly I cannot see the United States or the emerging economies of China and India cutting back in time, and they are the main source of emissions. The worst will happen and survivors will have to adapt to a hell of a climate.

PRECEDING PAGES: Monitoring Ozone in the atmosphere above the South Pole. Remote sensing from space now plays a vital part in assessing the past, present and future health of our planet.

There is a much more hopeful example of how things can change for the better through recognition of an environmental problem and concerted international efforts to deal with it. Chemicals called chlorofluorocarbons (CFCs) have been widely used in domestic sprays, refrigeration, air conditioning, and other industrial processes. However, scientists argued that the use of these was affecting the atmosphere, as when they were released they broke down into damaging elements, especially chlorine. The Earth has a layer of ozone gas high in the stratosphere, which protects us from exposure to dangerous levels of ultraviolet rays from the sun. But in 1985 a large hole in the ozone layer over Antarctica was seen from satellite images. It was evident that CFCs were primarily responsible for this ozone depletion, and that ultraviolet penetration was increasing as a result. For once, the world's nations came together quickly to create two global treaties by 1989, to which nearly 200 countries are now signatories. As a result, the production and use of CFCs is being phased out with the development of alternatives, and the recycling of existing sources such as old refrigerators is under tight control. The ozone layer is recovering, although it is being carefully monitored for any signs of further damage. Of course this success was a case where lines of action were clear, and where the most influential nations pulled together for the common good. In the present situation the US government, in contrast, appears to be in denial about the reality of global warming and our role in it, preferring to talk about long-term technological solutions to any problems. President George W. Bush's chief climate adviser, James Connaughton, has said he does not believe anyone could forecast a safe level of greenhouse gas emissions, and cutting them could harm the world economy. But there is hope that this short-term thinking will change for the better with Bush's departure, and there is now a groundswell of scientific and popular opinion in America about the scale of the problem and the need to address it.

If we look back at our early history it is clear that human populations only ever had a tenuous grip on their homelands. In our ancestral home of Africa, the main threat came from changes in rainfall and vegetation, and the knock-on effects these had on water and food. In northern Europe, the dominant factor would have been temperature, and particularly winter temperature. As we have seen, Britain was colonized by human populations at least eight times in 700,000 years, but seven of those were ultimately unsuccessful in the face of severe climate change, with only the occupation beginning about 11,500 years ago continuing to the present day. On the other side of the inhabited world, on another island, a strange and diminutive human species had by then come to the end of its long life. The ancestors of *Homo floresiensis*, nicknamed the 'hobbit', had apparently settled the island of Flores in Indonesia at least 800,000 years ago, and this species then evolved in splendid isolation in a tropical environment, far from the climatic vicissitudes of northern Europe. Yet that stability was wrecked by something completely different from climate change – a massive volcanic eruption 12,000 years ago devastated the whole island and seemingly brought this fascinating experiment in human evolution to an end. So contingency (chance) has always been a major factor in shaping the course of human evolution. The difference now is that we are largely responsible for the major uncertainties facing our species and life on the planet.

Looking at human history, which populations have coped best with rapid climatic or environmental change, and how did they do it? The species *Homo erectus* lasted over a million years, putting our recent emergence and 'success'

BELOW: Keli Mutu volcano on the island of Flores, in Indonesia. Flores lies in a highly active volcanic region, and a massive eruption about 12,000 years ago may have caused the extinction of dwarf forms of the elephant-like *Stegodon* and of ancient humans (*Homo floresiensis* – the 'Hobbit').

in the shade by comparison. But that longevity was achieved by dispersal and diversification in small numbers across a wide swathe of the tropics and subtropics of the Old World, such that climatic and environmental insults would only ever hit part of the human population rather than all of it. Our much larger numbers and the threat of major climate change across the whole planet means that we are in a very different situation.

The first people we know to have persisted through major changes in conditions were the Neanderthals, and we find them alongside remains of hippos and elephants in Italy, and mammoth and reindeer in Britain. They were highly mobile and moved across the landscape in small numbers to exploit a range of plant and animal resources. But when the climate changed rapidly it seems they found adapting quickly enough a problem, and when times were hard, they retreated to sheltered and more stable refugia in regions like the Dordogne and southern Europe, bouncing back if and when things improved.

Some of the Cro-Magnons fared better, and the Gravettians of 27,000 years ago were able to survive in harsher climatic conditions than the Neanderthals. They seem to have been more flexible and able to switch mobility patterns, at times settling in large camps with stable supplies, at other times dispersing and diversifying to gather scattered and varied food resources. They evidently had good buffering from extreme weather in the form of efficient fires, sewn skin clothing and tents, and the addition of nets to trap small game may have allowed the whole community – men, women, young and old – to participate fully in hunting activities for the first time. But the extensive spread of so-called Venus figurines at this time across Europe and Asia, as far east as Lake Baikal, gives a further clue to the success of these Cro-Magnons: their wide social networks. While Neanderthal social groups probably stretched their contacts over miles, through sending social messages in the forms of symbols and trade, the Cro-Magnons straddled continents. This gave them the potential to exchange mates, information, resources and ideas over wide areas and increased their effective group sizes enormously, all important when local resources were unpredictable and it might be necessary at times to rely on the support of neighbours to bale you out.

Of course the Gravettians in Europe probably only ever numbered in tens of thousands, rather than millions, and they only had to worry about crossing social and geographical, rather than political, barriers. However, the importance of cooperation rather than conflict in challenging times is a lesson for Europe and the world today. As stress builds up in the face of serious environmental changes, the continent could go in two different directions, following examples 27,000 years apart: Gravettian (flexibility and cooperation) or Balkan (factionalism and conflict). If we do not cooperate for the common good, we will certainly sink separately (in some cases literally, as sea levels rise). If we can face the challenges of global warming together, we can look to a new and more hopeful future. Britain was joined to Europe for most of its prehistory, and if sea levels rise it will become even more isolated physically. But in terms of a common future, we must extend our social networks of shared purpose a lot further than did the Gravettians who buried the Red Lady of Paviland. The future of *Homo sapiens* across the whole world is now as precarious as *Homo britannicus* ever was in that small peninsula of Europe called Britain, and the destiny of the whole world, not just our survival, depends on us.

OVERLEAF: Brighton Beach, 19 July 2006: British *Homo sapiens* enjoying the hottest July day on record. Even hotter days are expected as global warming takes full effect.

While it may be true that human kind cannot bear very much reality, one hopeful sign is the way the debate about climate change now has a very high profile. The shocking events of 9/11 happened the month before AHOB began its work in 2001, and my reaction to it then (on the Edge website, which debates scientific issues) was concern about the effect it would have on global priorities: 'I . . . fear that the world will forget the even greater threat we all face from global warming. If we do not start to face up to this threat properly, the chaos that will ensue over the next century as half the Earth tries to relocate to find food and water will make these recent events, awful as they are, pale into insignificance.' When I began to plan this book two years later, one of my main objectives was to bring the reality of future climate change, and our central role in it, to public attention. Things have changed for the better in the last four years, if not in significant actions to really address the threat, at least in a general recognition of the immense scale of the problems we face. Scientists, politicians, rock stars and media figures like David Attenborough are all prepared to acknowledge the reality, but it is now up to every one of us to do something about it if humans everywhere are not to share the fate of so many generations of *Homo britannicus* who were unable to cope with their changing world.

FREE ADMISSION & DECK CHAIRS FREE
Ice Tea

BRIGHTON PIER
Ice Tea

APPENDIX
THE AHOB TEAM

In the book I have regularly referred to 'AHOB members' and 'AHOB research' without giving details of the team that carried out the work. In fact the project has fourteen members who collaborate closely and meet regularly to discuss AHOB and its progress, and sixteen associate members who form a further collaborative network with the AHOB team. This appendix, prepared from interviews conducted by the science writer Sarah Lazarus, gives AHOB members (and one associate member, as an example) the opportunity to talk in their own words about AHOB and their role in the project, starting with me.

CHRIS STRINGER

I'm a research leader in Human Origins at the Natural History Museum, London, Director of AHOB, and the author of this book.

My interest in human evolution started at primary school. I was fascinated by fossils, and at the age of nine or ten did a school project on Neanderthals. My interest grew throughout my school years but I had no idea that you could actually work in this area, so I planned to study medicine. I had a place at medical school lined up, when I chanced upon University College London's prospectus. It was arranged alphabetically, and Anthropology was on the first page.

The course offered archaeology, human evolution, genetics and social anthropology. Suddenly medicine seemed much less appealing. So I phoned the college, was invited for an interview, and offered a place. Much to the amazement of my teachers and parents, I dropped medicine and took up this subject, which I had only just learnt existed.

I finished my degree in 1969 and wanted to carry on studying human evolution, but in those days you needed the best possible degree to get a grant. In the meantime I was offered a temporary job at the Natural History Museum by the then head of Anthropology, Don Brothwell, whom I had met whilst visiting to look at their fossil collections. The job lasted nine months after which, in the absence of further opportunities, I was planning to do teacher training. But a couple of weeks before I was due to start, Don got a call from the University of Bristol. They had spare funding for a PhD for someone working on human evolution. Don recommended me, so I abandoned the teaching idea and instead took up this opportunity in Bristol, where I examined the relationship between Neanderthals and modern humans in order to assess whether or not Neanderthals were reasonable ancestors for our own species. The data I accumulated strongly suggested they were not. They were too different – they had their own evolutionary line and had developed their own specializations.

After my PhD it was touch and go again, but something else came up trumps. A post for a human evolution researcher was advertised at the Natural History Museum. Don Brothwell offered me an interview, I got the job, and I've never left. I'm now a research leader in Human Origins, an informal group of anthropologists based in the Department of Palaeontology. My research covers human evolution over the last million years. I've done a lot more work suggesting that Neanderthals were not our ancestors, and from about 1984 I was drawn into the developing 'Out of Africa' debate, and I've been in the middle of that ever since.

Andy Currant and I have worked together at the Museum since I arrived here in 1973, and our first joint project was to excavate caves in the Gower Peninsula. In the 1980s we started working with Roger Jacobi at Gough's Cave at Cheddar, and then in the 1990s Boxgrove became a focus of research. In 1993 the human tibia turned up and I started studying that with Simon Parfitt. So my collaborative network gradually expanded, and by the time AHOB came around, many of my working relationships had been established for years. In 1996 I joined a new Cambridge-based initiative called the Stage 3 Project. This looked at the period when Neanderthals died out and modern humans took over in Europe (between 30,000 and 50,000 years ago). The Stage 3 strategy was to get a team of scientists from different disciplines to study every aspect of the problem. There were plant, animal, human evolution and geology specialists amongst others. That experience showed me how a team could bring different strengths and knowledge to a project, and achieve something that was much greater than the sum of its individual parts.

Around that time, a series of intriguing questions about the British Palaeolithic started cropping up. Did Boxgrove really represent the earliest human occupation in Britain? What was the significance of all the evidence from the Hoxnian interglacial? Could we resolve the Clactonian/Acheulean puzzle, or find out when and why the Neanderthals went extinct? It was clear that the British Palaeolithic was ripe for something like the Stage 3 approach. But what was needed was funding. Then in 1999, Leverhulme – a charitable trust with a long history of supporting scientific endeavour – announced a new initiative offering grants for research focusing on early human settlement of a particular region. The sums offered were much larger than usual, but the brief was very general. There were rumours that the focus did not extend to the Palaeolithic. Nonetheless, we took a gamble and applied.

During the application process, I started to build a team. The first to join me were Andy, Roger and Simon Parfitt because I was already working with them. With all the fantastic collections at the British Museum, Nick Ashton was an obvious candidate too. While the Natural History Museum has an unparalleled collection of Britain's Ice Age mammals, the British Museum looks after most of the archaeology, much of which comes from the same sites. Fossil mammal specialists and archaeologists head to the respective museums to research the collections, but no one had ever studied them together comprehensively. We realized that by combining the collections and investigating the complete story, we could learn so much more. We designed the project so that it would address seven major questions. To answer them, we proposed a detailed re-examination of existing collections, and some new excavations as well. A key benefit of the funding was that it gave Roger and Simon salaries for five years, enabling them to focus on critical research.

From there, we gradually built up the team from further networks of collaborators, inviting experts on such diverse topics as palaeoclimate, geology, and stable isotope analysis to get involved. As the director of AHOB, I'm nominally responsible for all the administration, although thankfully I'm able to share the work around. I have to send Leverhulme the annual accounts and progress report, the latter assembled by the whole group with the invaluable support of the team of administrators here at the Museum. I also have to make sure that the project keeps an eye on its goals and that people don't become too distracted and head down research avenues that might not be so productive, or would lead us away from our core objectives. The whole team meets four times a year to discuss our achievements and plans for the future, and we also have workshops to discuss the science. With such a large group, opinions often vary tremendously. Dating is quite a controversial issue – it's not an exact science and different people have strong method preferences. Ultimately, it's down to me to get people to pull in the same direction so we can move forwards. Sometimes, if we're putting a paper together, it can be a real challenge to find a consensus view to which everyone is happy to put his or her names.

For my own work on the project I've visited most of the AHOB sites, although it's fair to say that I get much more excited by those where human bones have been discovered. I've done considerable work on Pontnewydd, a Welsh cave site. Early Neanderthal teeth from about 200,000 years ago were recovered some time back, but are only now being studied in detail. I've also collaborated on work to estimate the age at death of the individual whose tibia was found at Boxgrove. Bone studies revealed that the man could have reached 40, a decent age at that time. I have worked with Mike Richards on using isotope analyses to understand ancient diets, and with Roger and Andy on getting more accurate dates for sites between 130,000 and 70,000 years ago, a critical period when people seem to have been absent from Britain. And I have continued my work on the Swanscombe skull, comparing it with contemporary fossils from other parts of Europe. We need to ascertain if one or more different populations inhabited Britain during this time. Swanscombe is a key site for the Clactonian/Acheulean debate, so more information will be of great value.

I'm especially fascinated by the early sites in East Anglia and what they tell us about early human occupation. Working at Pakefield has been an extraordinary experience. Of all the outputs of the project, I suspect this will have the biggest impact in terms of media and public interest, and will be what the project is most remembered for. We're pushing back the first arrival of humans in Britain by 200,000 years and that's big news. We've also been able to reconstruct the environment in which these people lived because the plant and animal evidence is so rich. The Pakefield discovery was published in the journal *Nature* in 2005 with a long list of names on the paper – AHOB members, associates and collaborators, including several dedicated local workers. This reveals the strength of the project: with so many people working together we can create the most detailed picture possible. A small team could never have achieved this.

Roger's paper on the artefacts from Gough's Cave, Cheddar, for which he won an award, is also a tremendous piece of work. While Pakefield shows the breadth of the project, with so many people approaching the subject from different perspectives, Roger's paper illustrates the depth. It shows how much one person can achieve when they're given the time and resources they need to concentrate on a particular topic. I hope AHOB will inspire other teams to adopt a similar multidisciplinary approach. Our experience shows how much you can achieve this way. We've significantly advanced research into the British Pleistocene and Palaeolithic, and put a wealth of information in the public domain. Before AHOB, both the faunal and archaeological records needed attention. We've filtered through the evidence, standardized the records, and created a database that is quality-controlled and sets out the best possible summation of our knowledge. People know that some of the country's leading experts produced these records and that they can be trusted. Our success is due, in part, to the strength of our team, and in part to being in the right place at the right time.

Following substantial coastal erosion in East Anglia, sites that had been buried for a hundred years have suddenly been exposed and we've been able to take advantage of that. I've learned so much from the project. The other team members have taught me about a diverse range of subjects, and it's been enlightening to see how people address the same issue from such different perspectives. We're now finishing AHOB and can assess what progress we have made, and start to plan our future research. And the great news is that the success of AHOB has meant that the Leverhulme Trust is awarding us a further large grant to continue the most productive collaborations and research. In the meantime, I'm looking forward to the last year of the project. A key goal for 2006–7 is to pull all our research together and publish a conference volume and a monograph. There may also be a touring exhibition of our findings, and I hope this book will be considered another successful output of the project.

ANDY CURRANT

When I began to develop the AHOB Project in the year 1999, Andy Currant was the first person I wanted in the team. He has been a colleague since I joined the Museum and we have dug together at many sites in England, Wales and Gibraltar. As both a researcher and the Curator of Fossil Mammals, he has an unrivalled knowledge of the Museum's fossil mammal collections. His warm personality and interpersonal skills are also invaluable during difficult times.

I'm the Collections Manager for Vertebrates and Anthropology at the Natural History Museum. I mainly focus on the Ice Age mammal bones. I've been at the Museum for thirty-four years and have come to know our collections exceedingly well. I've also done a phenomenal amount of fieldwork – when I was younger and thinner I was quite handy at exploring narrow caves and squeezing through small holes. I've been digging since the age of four. My great-aunt Hilda lived in St Albans and had connections with a dig at Verulamium nearby. I got to help out, scrubbing bits of pot and bone. I loved it and became passionately interested in stuff in the ground (my parents would say that I never looked up), spending my time digging holes all over the garden and finding all sorts of wonderful things. By the time I left school I had a lot of digging experience. I was due to study at Sheffield University but then my mother noticed an advert for a job at the Natural History Museum so I applied for that instead. I started off as a vermin-grade curator and then gradually floated up the system to my present position.

When in 1999 Chris learned that the Leverhulme Trust were interested in funding a major research programme, I put together much of the original AHOB draft application. People have been collecting human artefacts and fossil mammals for over 250 years and collections have ended up all over the place – in small museums, big museums, private collections, on mantelpieces, in cellars and under beds. My wish was to re-integrate, on paper, those long-separated collections, to see what we could say about human activity through time in the British Isles. However, we realized that the project needed a sexy front end involving original research. One topic we picked was stable isotope analysis, as this area seemed to offer huge potential. Although that particular part of the project is still being delivered, the discoveries we made while knitting together collections have exceeded any of our expectations.

Other revelations have come about through fieldwork. Some really important sites such as Pakefield, Happisburgh and Hoxne have become available in the course of the project. It's been phenomenal, as if someone opened the celestial letterbox and dropped these fantastic sites into our laps just when we had the resources to deal with them.

One of my key roles has been as a facilitator – going through the collections, re-identifying bones, looking for cut marks and other signs of human activity, and drawing people's attention to them. For my own research I've worked closely with Roger Jacobi to assemble and provide mammalian data on the Late Pleistocene. Roger's a top-notch archaeologist but he also shares my enthusiasm for fossil mammals. From early on we noticed that groups of mammals are repeated at different locations. Some sites have bison and reindeer, while others feature woolly mammoth, woolly rhino, hyaena and horse; the patterns are very consistent. The study of Ice Ages used to be dominated by botanists and their pollen assemblages, and what was then known about mammal assemblages didn't fit in with the pollen-based climatic record. What we've done is to create a mammal-based framework for the Quaternary.

AHOB gave us the opportunity to fit the human artefacts back into our mammal-based model. This led us to the extraordinary discovery that humans first came to Britain earlier than anyone had ever thought. We also confirmed that human occupation wasn't continuous. Roger and I have collaborated to try to understand the long period after about 170,000 years ago when humans seem to be absent from Britain. About 80,000 years ago there were lots of game animals like bison and reindeer in Britain, so why weren't people coming to hunt them? Maybe there were terrible flies, maybe the contemporary brown bear, the size of a polar bear, was sufficiently terrifying to keep people out, or maybe Britain was an island then. We've also tried to pinpoint exactly when people came back to Britain during the middle of the last Ice Age and what they then did. It's a job that's made more challenging because spotted hyaenas seem to have eaten a lot of the evidence.

One of my approaches is to look at anomalies in the fossil record and reassess previous theories. I've recently been working on material from Kent's Cavern in Devon where, in 1826, Reverend John MacEnery found some sabre-toothed cat canines. These have literally been bones of contention ever since. They were found in a late Pleistocene context, associated with human artefacts, but people didn't believe the first-hand observations of the excavators and insisted that the teeth were at least 500,000 years old. I'm sure they were wrong. As part of AHOB we've arranged to have the teeth dated and I'm looking forward to being proved right! This was a critical period for humans in north-west Europe. The Neanderthals were on the way out and anatomically modern humans were taking over. The hyaenas were unbelievably abundant. Given that a modern hyaena, a smaller animal, will eat the tyres off a Landrover, we can assume they caused huge environmental stress. Until recently we didn't realize that there was an additional predator, the sabre-toothed cat, around at the same time. The extra pressure it exerted could have tipped the balance in favour of modern humans. The Neanderthals may have been less able to protect themselves, and in these circumstances a small margin can make a big difference.

Throughout my five years working on AHOB, I've tried to keep it on the straight and narrow by occasionally reminding fellow contributors what our primary aims are. I've also acted as a general-purpose trouble-shooter from time to time – diplomacy is sometimes necessary even in the best regulated of households! I've had huge fun working on the project. It's brought together different threads of data stretching back throughout my career. A key principle of ours is to avoid being blinkered by preconceptions and to try to look at things with fresh eyes. Sometimes even things you have dug up yourself can come back and surprise you. We've accumulated an extraordinary wealth of verified historical data along with some new interpretations that we will now be able to put into the public domain. I hope it will provide a useful tool for the next generation of researchers.

ROGER JACOBI

I have known Roger Jacobi nearly as long as I have known Andy, and we three worked together for several years at Gough's Cave in Cheddar. He is one of Britain's foremost experts on Upper Palaeolithic stone tools and is often to be found travelling round the country, scouring museums and re-evaluating collections. His award-winning paper on Gough's Cave is one of the prize outputs of the AHOB project.

I'm based in the Department of Prehistory and Europe at the British Museum, where I work mainly with stones, but also sometimes with bones. My speciality is Palaeolithic and Mesolithic stone artefacts, but I frequently collaborate with Andy Currant to investigate Late Pleistocene fauna. I've worked with Andy and Chris Stringer for almost thirty years. We've done many excavations together and we're good friends. I got involved in AHOB because Chris caught me lurking in the Natural History Museum one afternoon, and asked me to join the team. I think he recruited me mainly for what's in my head – because of all my travelling I've probably seen more of Britain's late Pleistocene artefacts than anyone else.

I'm the most itinerant member of the project. Many of my colleagues seem to work in East Anglia and I'm the one who treks off to the north of England and Wales. I've always been a traveller. If you want to really understand the archaeology, you've got to go and see the collections for yourself. They are often widely dispersed, but by seeing and recording the material I can reunite them, at least on paper. I've recently worked on Church Hole, the first British site to produce Palaeolithic wall art. The finds from that site, which was excavated in the 1870s, are scattered around ten different museums. While some AHOB members are questioning the past by doing new excavations, others, like Andy and myself, are re-examining and revivifying old collections.

Britain's history of collecting extends over more than 200 years and is uniquely impressive in terms of what we've collected and preserved and the information we've recorded alongside the material. During the nineteenth century much money was spent establishing public museums, particularly in the wealthy northern towns. When I visit those museums, I might be looking for something specific or I might simply be browsing. Sometimes I'm amazed at the treasures I find. As well as objects I also consult everything that's ever been written about the sites – notebooks, letters and diaries. My golden rule for this kind of detective work is: no matter what, photocopy it, you might never see it again. If you're really methodical about it, seemingly disparate scraps of paper eventually piece together to provide useful information.

Pin Hole Cave, the key site for reconstructing the archaeology of Creswell Crags, has a particularly intriguing history. It was excavated, in the 1920s and 1930s, by an amateur archaeologist called Leslie Armstrong. He was a bit of a loner and was quite badly ostracized by some members of the academic establishment, but he did a superb job at Pin Hole. The cave's archaeology has never been properly written up so that's something I'm planning to do. My research has involved re-identifying the fauna, dating parts of it, and drawing and recording all the Stone Age archaeology. The cave contains both Neanderthal and modern human stone tools, while the fauna represents one of the largest collections from the middle part of the late Pleistocene in Britain. It appears that the modern humans used the cave more intensively than the Neanderthals, and trapped Arctic hares.

The richest site of all is Gough's Cave, a show cave at Cheddar where I've been working for more than twenty-five years. From the 1890s, developers gradually lowered the floor, to facilitate access for tourists, and in doing so revealed late Ice Age implements and animal remains. During a big push to get the cave open in the late 1920s a huge quantity of archaeology and animal

bones, dating to around 14,000 years ago, was discovered. Fortunately the estate's agent knew a bit about cave excavation and they rescued a large and important collection of material. Items continued to be found and recorded until 1986 when I spotted human ribs, teeth and implements sticking out of a newly cleared section. Realizing the importance of the site, Chris, Andy and I continued to excavate there for several years after that.

We found an amazing collection of human bones and this led to the cave being dubbed the Cannibal Cave. The evidence is pretty convincing. The bones represent five individuals, from a child about four years old up to young adults. We found them jumbled up with animal food debris, flint implements and assorted domestic rubbish. They bear the same cut marks as the animal bones and look as if they were broken up to access the bone marrow. It is hard not to conclude that, along with horses and red deer, those people formed part of the food chain. Interestingly, the Gough's Cave inhabitants used tools made from mammoth ivory and reindeer antler, but neither of those animals lived near Cheddar then, so they were either encountering those animals elsewhere or perhaps trading in raw materials.

Despite over a century's worth of collection and twenty-five years of my own research, a comprehensive account of the archaeology of Gough's Cave had not been published until the AHOB project came along. I spent about two years compiling a detailed account of the stone tools and setting them into the context of late Ice Age life at Cheddar. The paper took so long to produce partly because all the tools had to be illustrated. Nobs (Robert Symmons), then AHOB's meticulous research assistant, has been a tremendous help. I'm not a very practical person – I don't drive and I don't use computers – so I relied on him enormously. He's absolutely brilliant at computer graphics, so we've worked very well together. I've done similar work at Kent's Cavern, near Torquay, another site that people have been picking at for years. John MacEnery initially excavated the cave in the 1820s. Many of the things that puzzled him now make sense. MacEnery didn't have the knowledge we have today but he made a point of noting everything and the information he left has proved critical. Following MacEnery, the cave was excavated by a local geologist called William Pengelly. He produced a fantastic grid of the cave so all the objects he found could be replaced *in situ*. I'm intending to plot the Pengelly material into a computer plan of the cave produced by two American scientists who based their work on his records.

In order to date human activity, wherever possible, we select cut or modified bones or tools. Until recently the most commonly available materials – bone, teeth, ivory and antler – were also the most problematic for dating. The samples need careful pre-treatment to remove recent contamination and the technology we used was not sufficiently effective. Recently the Research Laboratory for Archaeology at Oxford, who run all our samples, developed new, improved pre-treatment methods. So the radiocarbon dates we've been getting on AHOB are probably more reliable than ever before. Some of the shifts in age have been really dramatic.

Our work for AHOB is really hands-on and this gives it tremendous value – if you don't know the material first hand, you can't properly evaluate it. It's painstaking work. You might spend a week looking through a faunal collection and find only a single cut mark, but even that makes it worthwhile. In fact, it's the only way to do it. Other researchers who base their work on our findings will know that we've sought all the traceable objects in a collection, looked at them, and dated them with the best methods available. It's quality-controlled. AHOB is putting Britain's Pleistocene archaeology into sharper focus. We're lucky to be working on it right now because it's possible to apply all sorts of new methods and thinking and ideas to the material. The project came along at the perfect time.

SIMON PARFITT

I got to know Simon Parfitt through our collaboration on the Boxgrove site, and we have worked together on describing the human shinbone and teeth from there. He is one of AHOB's faunal specialists, and with his unmatched enthusiasm for fossils of tiny creatures like voles and shrews he's been able to reveal a wealth of information about the ancient environmental and human history of Britain.

I discovered the joys of archaeological excavations at the age of 13 when I joined a dig at a Roman temple site. I spent every subsequent school summer holiday as a volunteer helping archaeologists with their work. I had fantastic experiences and relished the freedom of living away from home, surrounded by interesting people. At the temple we found the remains of sacrificial sheep, goats and fowl in such huge quantities that we had to use wheelbarrows to cart them away. The animal bones fascinated me and so began a lifelong interest in vertebrate palaeontology. It's a bit like detective work because you have to piece together the clues to reconstruct the scene. After school I enrolled at the Institute of Archaeology in London. We had a brilliant, inspirational lecturer called Don Brothwell who sparked my interest in environmental history. In my final year I heard about an excavation starting at Boxgrove, and managed to get a job helping with the animal bones. It was both exciting and daunting because it's such an important site. During that summer I realized that this was what I wanted to do with my life. Starting in 1985 I spent six months of every year, for over a decade, doing fieldwork at Boxgrove. The other six months were spent at the Natural History Museum sorting, identifying, conserving and cataloguing everything we collected. In 1993 I identified a human tibia bone. The excavation was, by then, coming to an end, and we had given up all hope of finding a human fossil so it was incredibly exciting. At the time, this represented (along with a jawbone from Germany) the earliest fossil human in all of Europe.

So although my academic background was in archaeology, my expertise is in animal remains. I find them compelling because they can provide so much information: whether the environment was warm or cold, whether there were trees or rivers, what people ate and how they captured and butchered their prey. We also use animal remains to date archaeological sites because the evolution and extinction of some species, such as voles, are well documented. Voles, along with shrews and mice, are the kind of small creatures that I particularly like working on. Most researchers ignore them in favour of the big, more obviously appealing animals like woolly mammoths, so this field is virtually unexplored and is wide open for new research. In many ways the smaller animals are more interesting, and because they can be found in abundance they provide a lot more information than their larger counterparts.

Boxgrove is a prime example of the usefulness of animal material. The beautiful hand axes found there had led researchers to believe that the site was relatively young, comparable to Swanscombe at around 400,000 years old. But our analysis of the animal bones revealed that it is in fact about 100,000 years older and that humans lived there before, rather than after, the Anglian glaciation.

I first got involved with the AHOB project through my association with Chris Stringer at the Natural History Museum. When we started work the Boxgrove site, dated at about 500,000 years, provided the earliest unambiguous evidence of human habitation. We embarked on a calculated search for earlier archaeology but we could never have predicted that we'd make the amazing finds we did. We headed to East Anglia. The coastline there was notoriously devoid of archaeology and this had never made sense. The animal remains and local geology indicated that this area, like the Boxgrove region, was scored by river systems and enjoyed a warm climate. So where were the people? I searched for evidence in the Natural History Museum's extensive collection. Eventually, while examining fossils that had been found more than a hundred years ago, I discovered a bison bone with clear cut marks on it. It had been discovered in the 'Forest Bed' at Happisburgh in Norfolk, but the precise details of its location were not documented. Then, in 2000, an amateur archaeologist found a handaxe. Crucially, it was buried in sediments, so it could be dated. These finds provided the first conclusive evidence of human occupation of the area dating to perhaps half a million years ago.

We then conducted numerous excavations along a large stretch of the East Anglian coast, from north Suffolk all the way round to The Wash. The big breakthrough was at Pakefield. The artefacts found there pre-dated Boxgrove by as much as a quarter of a million years, pushing the colonization of Britain back to an earlier time than anyone had anticipated. In fact, it's the earliest known site in northern Europe. It's incredibly exciting. We identified thirty-two flakes of flint at Pakefield but no handaxes. Opinions differ as to the significance of this, and the evidence doesn't present a clear pattern. It's possible that the people knew how to make only simple tools, but it could be that they were limited by the available raw materials. My feeling is that different groups, arriving in Britain from various parts of Europe, brought different tool kits with them. Excavating at Pakefield was hair-raising. We had to work kneeling at the foot of a towering sandy cliff with huge blocks of clay above our heads that threatened to fall off at any moment. We excavated with trowels and filled about fifty sacks. Robert Symmons then spent months sifting through the samples, picking out tiny chips of flint. These minute artefacts allow us to envisage someone sitting on gravel, probably near a river bank, prospecting for workable pieces of flint. They would then bash the flint to remove flakes, leaving the scattered chips that we recovered 700,000 years later.

As well as flint chips the samples contained hundreds of small bone fragments and it was my job to look at those. There were numerous different animals – fish, amphibians, birds, reptiles and small mammals. Amongst the most interesting specimens were a bat tooth, a squirrel tooth (indicating that the landscape was wooded in parts), and two teeth from an extremely rare water shrew known from only twelve other sites.

One of the key aims of the AHOB project is the investigation of numerous archaeological sites such as Pakefield, High Lodge, Westbury Cave and Warren Hill, which belong to the early time period before the Anglian glaciation. The sites represent a wide range of climates and human technologies and we're now working to fit them into a chronological sequence.

The AHOB project is unique because it brought together a whole range of approaches to the subject. For me, it provided an amazing opportunity to both explore new sites and revisit old collections in order to unearth evidence and establish connections.

JIM ROSE

Jim Rose is a long-term colleague and one of AHOB's geomorphologists (geomorphology is the study of landforms). He is a major figure in studies of the landscape of Britain over the last 3 million years and made his name with some groundbreaking ideas on how this landscape has changed. He is a Professor of Geography at the University of London, and a leading international figure in Quaternary science.

Some of my most significant contributions to AHOB stem from my earlier achievements. Some of these are what might be called 'startling realizations' that have explained hitherto unresolved problems. These revelations have greatly added to our understanding of Britain's recent geological history.

The first of these 'blinding flashes' occurred in the mid 1980s when I was studying glacial deposits

in eastern England. A Lincolnshire sand and gravel company had asked me to find them some new deposits. While studying the geology around their quarry, I realized I was standing on the bed of a major ancient river that had nothing to do with the glaciation of the area! Having found the company some more sand and gravel, I then used my knowledge of the Quaternary geology of midland and eastern England to trace the ancient river's course. I discovered that it had been a very large watercourse and had drained from near Stratford upon Avon to the North Sea near Lowestoft, but that it no longer formed part of the landscape. It is a buried river system and indeed forms the 'Great Lost River of England'. I named it the Bytham River, after the village of Castle Bytham in south Lincolnshire, which is close to where I made the discovery. The Bytham River turned out to be of tremendous importance, geologically and archaeologically. Geologically, it demonstrated that deposits considered to be evidence for the penultimate (Wolstonian) glaciation of Britain were in fact much older – something some people have still not recognized. Archaeologically, it provided a context for evidence of humans in Britain before the most extensive glaciation, the Anglian, which occurred about 450,000 years ago.

The second 'binding flash' occurred around 1990 while I was working in the field with my (then) research student, Simon Lewis (also a member of AHOB). We were studying the sediments at High Lodge in Suffolk, a famous and controversial archaeological site. Simon and I were hard at work when suddenly we stopped and looked up at one another. I said: 'these are Bytham deposits!' and everything fell into place. All subsequent work has confirmed this to be the case. The sediments at High Lodge are the overbank floodplain of the Bytham River where people lived around 500,000 years ago. Existing publications had recorded human artefacts here, but had not recognized the deposits as Bytham sands and gravels. While conducting research at other Bytham sites, I had found a number of possible human-struck flint flakes, but the stone tools from High Lodge provided the first unequivocal evidence of human habitation associated with this long-lost river system.

The third 'blinding flash' occurred while I was doing collaborative work with members of the British Geological Survey. We were at a Bytham River site in East Anglia and I was being teased by my colleagues who had recognized big glacial erratics in the Bytham sands and gravels that I had not previously seen. Initially this was very embarrassing, but then I forgot the embarrassment because it meant that here was evidence for a glaciation long before the Anglian glaciation that had destroyed the Bytham River! This completely overturned my own theories about Britain's glacial history. Having considered all the geological and geomorphological evidence, we proposed the term Happisburgh glaciation. The Happisburgh event occurred about 650,000 years ago. Many in Britain, including some members of the AHOB team, do not believe in the existence of this glaciation. They continue to defend my original theory, proposed in 1979, that a single glaciation – the Anglian – occurred during this time period. It is interesting that the Happisburgh glaciation did not completely destroy the Bytham River because it did not extend far enough south. By contrast the Anglian glaciation, which reached London, either eroded or buried the Bytham river valley.

Around the same time, a beautiful handaxe was discovered at Happisburgh, in East Anglia, beneath the glacial deposits. This indicates that the humans lived here at least 650,000 years ago. This is extremely old, but the geology at Pakefield near Lowestoft provides evidence for even more ancient archaeology that may date to 750,000 years BP. Ever since I started work on AHOB, I have argued that humans first came to Britain much earlier than most people believed. Originally, many on the team disagreed with me, but now we all share this opinion.

Over the last few years there have been some strong reactions to my ideas about a major pre-Anglian glaciation, and the presence of humans in Britain long before 500,000 BP. The alternative view is that the Happisburgh glaciation is the same age as the Anglian and that the archaelogy below the Happisburgh glacial deposits is roughly contemporary with the archaeology at Boxgrove, at around 500,000 years old. These beliefs are based largely on the biological evidence, which uses evolution and extinction as yardsticks. The assumption is that biological change (i.e. the first or last appearance of a species) takes place at the same time everywhere – if the change can be dated at one location, then it is thought safe to apply that date elsewhere. But we work within a fine resolution and this is unrealistic. There are numerous examples of species becoming locally extinct at different times, so I believe this is a flawed technique to use for dating. If it can be demonstrated that the changes took place at similar times, then I'm happy to use the biological method, but I cannot take it as an article of faith.

I have really enjoyed working as part of the AHOB team. There is immense value in collaborating with people from a variety of disciplines, with their different philosophies and prejudices. The existence of fundamental disagreements on major issues stimulates critical thinking and has contributed in no small way to the current belief that humans lived in Britain around 700,000 years ago. It is my opinion that this scientific debate that has been the project's biggest success.

DANIELLE SCHREVE

Danielle is an Ice Age mammal specialist, and has uncovered a wealth of information about the environments Britain's ancient humans lived in, as well as their eating habits. She has collaborated with various members of the group on a number of projects, including our publication of the Westbury site and sites in the Thames Valley.

I'm the third of the unholy trio of Ice Age mammal specialists working on AHOB. Along with Simon Parfitt and Andy Currant, I've concentrated on British fossil mammals to provide information about past environments, the age of sites, and their exploitation by early humans.

I couldn't have predicted that I would follow a scientific path when I left school with three A Levels in arts subjects. I started my degree at the Institute of Archaeology expecting to focus on classical topics, but soon discovered that early Stone Age archaeology and environments were even more compelling. I subsequently did a PhD on Ice Age mammals. Mammal remains can provide detailed information about when and how the animals lived. Every time Britain switched from an interglacial into a glacial period, the faunal slate would be wiped clean. It's like reshuffling the pack and dealing a completely different suite of animals with every major climatic shift. Andy Currant and I came up with the term Mammal Assemblage-Zone (MAZ) as a good way to describe these discrete groupings of mammals. I ploughed through the vast collections at the Natural History Museum, and conducted several new excavations, in order to define the MAZs. This became a key element in the AHOB project design as we used the mammal evidence as the basis for our chronology. When I'm investigating a site, I start by compiling a list of the species present. This tells me whether it was a warm or cold period and what the vegetation might have been like. I then consider, in more detail, whether species absences are genuine or simply an artefact of preservation. Some species are particularly good chronological markers. The presence of hippos, for example, indicates that the site is either older than 500,000 years or around 125,000 years. Absence of certain species can also give us concrete information – for example, the spotted hyaena was completely absent from Britain during the Hoxnian interglacial around 400,000 years ago.

I then focus on individual lineages and look at patterns of evolution. The water vole is an extremely informative animal. Over the last half a million years, water vole teeth have shown a progressive evolutionary trend. They grew in size while the thickest enamel on the biting surface moved from one side of the tooth to the other. By measuring the remains, the well-charted 'vole clock' allows us to place and order the sites in a chronological framework. We also see changes in size. Mammals adapt very quickly to environmental shifts, often by growing in size or dwarfing. Carnivores tend to get bigger in cold climates while herbivores sometimes shrink; during a particularly severe cold spell around 180,000 years ago horses, which started off with cart-horse dimensions, diminished until they were about the size of a Shetland pony. The first big assemblage I looked at was from Swanscombe. Other lines of evidence had suggested that Swanscombe was older than the famous Palaeolithic site at Hoxne. However, by comparing the mammals at the two sites, I worked out that they are roughly contemporary. They both pertain to the Hoxnian interglacial, during which the orbital configuration of the Earth and sun was the same as it is now. Environmental information from these sites is therefore relevant to questions about climate change today. Although concern focuses on global warming, we should also be aware that we might be due for another Ice Age. Most interglacials last about 10,000 years so we could be approaching the end of ours. The Hoxnian, however, is thought to have lasted around 20,000 years. If our interglacial proceeds in a similar fashion, we would have thousands of years of warm weather left. Our work for AHOB has revealed a clear pattern of development and human activity during the Hoxnian: the earlier part of the interglacial featured a more densely forested landscape but no handaxes; later the landscape became more open and handaxes appeared. Mark White and I believe that the evidence represents two waves of colonization of Britain. The first influx came from areas such as the north German plain, where the people did not make handaxes. The second wave came from Italy, France or Spain where handaxes were a well-established part of the local technological repertoire.

One of my most exciting AHOB projects has been working on the assemblage from a 60,000-year-old site at Lynford in Norfolk. An old river channel yielded an extraordinarily rich flint assemblage as well as tens of thousands of fossil bones, teeth and bone fragments, insects, plant remains and shells. The unusual density of animal remains strongly implied that Neanderthals had hunted and butchered large game there, so I investigated this theory. I found that the majority of bones and teeth were from woolly mammoths and was struck by the fact that virtually all of the meatiest limb bones were missing. Very few of the remains seemed to have been gnawed by carnivores, so I wondered whether the Neanderthals had taken them elsewhere. None of the bones bear cut marks, but mammoths had such thick hides and flesh that contact between bone and flint tool would rarely have been made. Even in experiments with modern elephant carcasses, experienced butchers leave no cut marks at all. A critical piece of evidence may be that the mammoth remains do not reflect the natural range of ages found in a modern elephant herd, where adult females and juveniles predominate. At Lynford, nearly 90 per cent of the specimens are from young to middle-aged adults and, where sex can be established, they are all male. This suggests a death assemblage pattern typically produced by human hunters. A healthy adult mammoth would have been dangerous to bring down but Neanderthals could have herded them towards cliff edges or trapped them in treacherous bogs. So although the only evidence is indirect, there are good grounds for speculating that Neanderthals were hunting mammoths. It appears that they arrived at the site 'tooled up' with handaxes they had manufactured elsewhere. And living in the harsh conditions of the last Ice Age, I can't see them passing up such a nutritious meal.

Working on AHOB has been fantastic. The project's strength lies in the range and diversity of experience of the team. We are accumulating an unparalleled dataset demonstrating how ancient British populations ebbed, flowed and evolved, and are gaining valuable experience in the use of novel techniques such as faunal and environmental stable isotope analysis. These methods provide an unprecedented level of detail, and have allowed us to test some of our crazier ideas about the age and environment of sites. It's inevitable that there are differences of opinion in the project but that's also part of the fun – it makes you think twice about the validity of your own evidence.

I am the only woman in the core AHOB team and although this does not make a jot of difference to the way we work together, I am aware of it as I look round the table at meetings. This reflects the more general paucity of female scientists in the UK. There are quite a number of female Quaternary palaeontologists on the Continent, but I can still count my female British colleagues on the fingers of one hand. I came from an arts background but with persistence, hard work and a few lucky breaks, I have been able to make it. Now that I'm a full-time academic, I hope I can encourage my female students to follow a scientific career.

NICK ASHTON

As Curator of the Department of the Prehistory of Europe at the British Museum, Nick Ashton brought vital knowledge of their immense collections to the AHOB project. He has organized many of our excavations and has spent countless hours in the field, investigating the enigmas of ancient human history.

Having been interested in history since a young age, I enrolled to study at the Institute of Archaeology. I planned to focus on medieval times but surprised myself by finding ancient archaeology more compelling. I leapfrogged the Romans, landed on the Neolithic, then drifted back in time to the Palaeolithic. I started work at the British Museum in 1983 and have been there ever since. My work focuses on the Lower and Middle Palaeolithic. My knowledge of the Museum's collections, which are vast, has been useful to the project – targeting the right artefacts provides us with the information we need and guides us towards sites for new excavations. I should add that although I know the collections fairly well, I can't compete with Roger Jacobi, whose brain is truly encyclopaedic. What motivates me is solving problems. Flints themselves aren't that exciting; but what they tell us about the mysteries of ancient human history is. It's the people that intrigue me, and the flints are a way to understand them.

Much of my fieldwork for AHOB has focused on the earliest human occupation of Britain. In the summer of

2004 I organized a three-week dig at Happisburgh. As well as being the site's archaeologist, I ran the dig on a day-to-day basis, contacting local landowners and council people, hiring equipment, organizing food and accommodation, roping in various specialists and co-ordinating everyone's work. This job was the most difficult logistical challenge I've ever faced. The site is rather inconveniently situated underneath a cliff on a beach and exposed only at low tide. We had to get a digging machine to come every day to clear away the sand deposited by the previous high tide. Because of varying tide times, we started work at 5 a.m. on some days and lunchtime on others. To dig out, sieve and process seven tonnes of material from the site was a considerable achievement.

Ever since a handaxe was found there in 2000, Happisburgh has been recognized as one of the earliest sites of human occupation in northern Europe. It's also important because there's a huge amount of environmental, plant and animal evidence directly associated with the archaeology, so we can re-create the landscape. The humans lived on the banks of a slow-flowing river, surrounded by coniferous woodland, in a climate that was slightly cooler than today's. Large herbivores such as elephant and rhino routinely trampled the river valley's vegetation, opening up the landscape. The people left numerous flint chips but few distinct knapping scatters, so it's likely they manufactured their handaxes elsewhere. Presumably they used their tools to butcher the abundant local game. A key objective of the dig was to look at the stratigraphy of the sediments in order to work out the dating. Some people, like Jim Rose, think the site is very early. The faunal evidence is more equivocal so it would be helpful to recover some more animal specimens.

I've also been greatly involved with work on the Hoxnian interglacial. The Hoxnian is particularly rich in sites providing good environmental data and archaeological evidence, so it's a fantastic period in which to closely observe human behaviour and patterns of habitation. The picture that's emerging indicates that people preferred to live by rivers rather than lakes. This makes sense: riverside environments are more dynamic, support diverse vegetation, and they are erosive environments so they provide access to flint, an essential raw material. Like lakes, they attract thirsty game. The river valleys also acted as corridors through a densely wooded landscape – the motorway system of ancient Britain, in fact. All the evidence points to a climate that was similar to or slightly warmer than today's. We get this climatic information from the presence of certain amphibians and reptiles that are very dependent on temperature, particularly the European pond terrapin. The terrapin doesn't brood its eggs but relies on the ambient temperature of the mud, which must reach at least 18°C in July, when the eggs hatch.

I've also waded into the Clactonian/Acheulean debate in the Hoxnian. The traditional view is that these two stone tool assemblages represent different groups of people with different technological cultures. I'm not so sure. I suspect that they may have been produced by the same group of people, undertaking different tasks at different places according to their needs and the availability of raw materials. Any assemblage is shaped by a plethora of contributing factors and I think you have to examine and unravel them all before making any assumptions. Whatever the answer, I'm sure the history is more complex than has previously been thought.

A later technology, the Levallois (Prepared Core), was introduced in Europe more or less synchronously, around 250,000 to 300,000 years ago. There are some who believe that it originated in Africa and was carried to Europe by a new species of hominid, but there's very little evidence to support this. I've examined a number of sites, including Purfleet in Essex, looking at tools that represent the genesis of Levallois and which refute that claim. I'm fascinated by what Levallois can tell us as the material expression of people's behaviours. Levallois was a more economical and mobile technology. Several serviceable flakes could be created from a single core. I think this technological advance developed in tandem with a range of social changes. The people were undergoing a process of 'neanderthalization' as they adapted to cooler, more open environments and became more proficient hunters. It's possible that social groups grew larger and that with their portable tools people cooperated to exploit the abundant herbivores over larger territories.

Another important question I've been working on is whether the creation of the English Channel might explain the period of human absence that started around 200,000 years ago. Collections from the River Thames terraces show that human populations gradually dwindled in the period leading up to the absence, which is surprising as you might expect population densities to increase over time. Prior to the creation of the Channel, Britain was connected to Europe. However, after the sea carved a channel through the chalk of what is now the Dover Strait, it was only during cold periods that sea-level dropped sufficiently to reconnect Britain to mainland Europe. At the moment we don't know the date of this event, but if it happened about 200,000 years ago, it might explain the dearth of humans in Britain after this time. There is, however, a site at Crayford that may overturn this theory. This is a fantastic site that was originally excavated in the 1880s, full of incredibly fresh artefacts that fit back together like a 3-dimensional jigsaw. At the moment, everyone disagrees on its dating. If, as I suspect, it dates to around 180,000 years ago, this would have a significant impact on our thinking about the period of absence.

AHOB has been a brilliant experience. A major part of my role has been in organizing and co-ordinating digs. I like bossing people around so this suited me perfectly! Having worked with many of the people involved before, I knew we'd all get along. By funding new field work, the project has given us the impetus to do lots of things that have only been talked about for ages. It has really pushed the research forward, with some astonishing results.

MARK WHITE

Mark White is another of AHOB's experts on Palaeolithic tools. His work on Neanderthal technology has much wider implications for our understanding of human behaviour and evolution.

I first did a degree in archaeology at University College London, then a PhD on British handaxe assemblages at Cambridge, and have been lecturing at Durham University since 1999. I specialize in Lower and Middle Palaeolithic stone tool technologies and human behaviour, and I've excavated at a large number of British Palaeolithic sites. At times my work feels less like archaeology and more like a minor international conflict.

I've been involved with AHOB since its conception, having worked with several other core members on field projects and papers over the years. It's been a great experience. Our work has pushed the British Middle Palaeolithic to the fore after decades of languishing in the dark. We seem to be getting a handle on the character and timing of the period.

I particularly like Neanderthals. What appeals to me about them is their otherness. They seem so like us at one level, and yet so different at another; it's like seeing a reflection but not quite recognizing it. I come from the school of thought that Neanderthals were more sophisticated in their behaviour than the archaeological record allows us to see. I think they were doing things to survive for which we have no material evidence. As archaeologists, we can almost never know about technologies based on organic materials. The differences between Neanderthals and modern humans would seem to be a question of degree, rather than kind.

I've spent a lot of time considering the questions raised by Levallois, a particular core-working method designed to produce flakes of a more or less predetermined form which became widespread during the Middle Palaeolithic. I started off by working with Roger Jacobi and Nick Ashton, documenting all the evidence we have in Britain. Rather than confining ourselves to the nitty-gritty of technology, we've also been looking more broadly at the changes in human population and behaviour that appear to accompany its development.

Some people have argued that Levallois was developed in Africa and introduced to Europe, but the evidence doesn't back them up in my opinion. Along with Nick Ashton, I have argued that it arose locally and quite independently in Europe and spread in a piecemeal way, disappearing and reappearing several times. The work we did on the material from the 300,000-year-old site at Purfleet, Essex, seems to support this. What you have here appears to be a proto-Levallois, where the rules of the method first start coming through.

In terms of their wider implications, the essential characteristics of Levallois cores are that predetermination and planning went into their manufacture, and that they are transportable sources of flakes, like a mobile tool production kit. Unlike the earlier Lower Palaeolithic handaxe, which moved as a tool, Levallois technology moves as potential for other tools. This might tell us something about how people were moving though their world and how they were organizing themselves. In a way, you might say that Levallois broke the tether between people and natural stone resources, liberating its makers to travel much greater distances without too great a risk.

I've also been investigating the Clactonian/ Acheulian issue. The stone tool assemblages from sites like Swanscombe, Little Thurrock and the eponymous Clacton do not include any handaxes. By contrast, many other Lower Palaeolithic British sites are rich in handaxes. These are referred to as Acheulean sites. The question is whether or not a real dichotomy exists. Nick and others have argued that the contrasting assemblages do not represent two separate cultures. They believe that the absence of handaxes at Clactonian sites might be

explained by the local availability of raw materials or because the people were engaged in different activities. They believe that the technical differences were not produced by different populations. I bought into this quite strongly until about seven years ago when I began to accept that the Clactonian did exist and should be recognized as a distinct culture. Whether the Acheulean developed from the Clactonian, or whether separate influxes of people arrived from different parts of Europe, I really don't know. I can't help but speculate, but haven't got any personally cherished solutions.

Like everyone in AHOB, I think humans were genuinely absent for extended periods from Britain. Various explanations have been proposed, but perhaps a more interesting question is why they came back when they did. A particularly intriguing example is the return represented at the Neanderthal site of Lynford, after a possible absence of 100,000 years. If the environmental reconstructions there are correct, then the climate was cold and unstable, with extremely harsh winters. The landscape, although rich in large fauna, was treeless and inhospitable. So why it appealed as a site for recolonization is far from clear.

The stone tools recovered at Lynford tell us a lot about the people who made them. The Neanderthals arrived at the site carrying handaxes that they had made elsewhere. Unlike the Lower Palaeolithic handaxe, which was more fixed, the edges of the Middle Palaeolithic ones were modified in subtly different ways. One edge might be fashioned as a scraper, another as a cutting edge, or they might have been notched for use as a spokeshave. There's also good evidence that the tips of handaxes frequently snapped off and were rejuvenated. People have often said that handaxes were like Swiss Army knives – I think this is questionable for the Lower Palaeolithic, but is definitely appropriate to Middle Palaeolithic ones.

As well as pre-manufactured tools, they probably carried blanks to enable them to produce more. They weren't, as was once thought, moving randomly around the landscape like hapless idiots waiting for chance hunting encounters or scavenging opportunities. Instead they prepared themselves, tooled up, and moved out into the landscape with the intention to hunt. Their handaxes, with their range of different functions, allowed them to respond to different situations as they arose. In this way Middle Palaeolithic handaxes have a lot in common with Levallois: they are a form of future proofing.

Obviously a key question concerning Neanderthals is what became of them. I think they were present in Europe in incredibly small numbers to start with – there were never wall-to-wall people in the Thames Valley. Such small numbers would be extremely susceptible to random population fluctuations, so they probably experienced local extinctions all the time. This would have been exacerbated by frequent major environmental upheavals when they routinely faced the kind of dramatic climate changes to which other animals respond by either leaving or going extinct. In terms of the final push, it seems likely that modern humans were implicated in some way. It's no coincidence that when modern humans left Africa and colonized the world, the rest of the planet's human species died off.

The AHOB project has been an enriching experience. The arguing that has gone on has been one its most productive aspects. What it has never become is some sort of mutual backslapping society – rather than praising each other's work, we sometimes can't actually agree on much at all. This has created a lively forum in which to ask questions and debate ideas. My own work on AHOB has generated more questions than answers. In terms of moving research forward, that's immensely beneficial.

DAVID POLLY

David Polly, a mammalian palaeontologist and pioneer of the Internet, was a perfect choice to help us create and develop AHOB's website and database. These provide information on the project for the public, and are an essential resource for the team members.

I specialize in mammal evolution and phylogeny, but my role on AHOB is very different – I'm in charge of the project's website and database.

Originally from the United States, I graduated from the University of Texas-Austin and then did a PhD at the University of California-Berkeley on the evolution of Creodonta, an extinct group of carnivorous mammals. During my PhD I started working with collection databases, Geographic Information Systems (GIS), and this new thing called the Internet. With a fellow student, Robert Guralnick, I developed one of the first hundred websites ever, the University of California Museum of Paleontology site. When the Internet first came to public attention during Al Gore's campaign to promote the Information Superhighway, the media really picked up on our site. Being full of pictures of dinosaurs, it, along with NASA's, was one of the few with widespread appeal. Most of the others were technical or science sites. So our profile boomed. We were featured on the cover of *Science*, on the official poster of the 2nd International WWW conference in 1994, and in numerous newspapers and magazines.

In 1996, my wife and I moved to London. During my first year, I worked as the web administrator for the Natural History Museum. I left in 1997 for a lectureship in Anatomy at Queen Mary, University of London, and then transferred to a lectureship in Evolutionary Biology in 2001.

My work combines quantitative genetics, developmental biology, and vertebrate palaeontology. I am interested in how the correlations between different parts of the skeleton and dentition influence the rate and direction of evolution. I am also interested in the effect of climate change on evolution, speciation, extinction and geographic distribution. To explore this, I compare living and fossil mammal populations. The Quaternary is particularly intriguing because many Ice Age species are still alive today, allowing palaeontological data to be compared directly with information from living animals.

My involvement in AHOB stems from my interest in the Quaternary and my experience with databases. In 1999, when I thought I might lose my position at Queen Mary through restructuring, I started looking for research funding and discovered a Wellcome Fellowship offering grants to study the effect on life of climate change. I hoped to compile, for the first time, a database of all the published information on British fossil mammal faunas. I approached Chris Stringer and Andy Currant to discuss my ideas, but they thought the published literature was too dated to make the project viable. Soon afterwards, the Leverhulme Trust announced their new grant programme on Early Human Settlement and an ultimately successful funding bid was launched, led by Chris. The funds they granted were sufficient to build up a small team that would re-evaluate the mammal faunas, their dating, and their association with human habitation. It also allowed for the creation of a database of that information. We all thought this was an excellent idea, and the result was the AHOB project.

So my main contribution to AHOB is running the website and co-ordinating the database. I'm not involved with digs or looking at the collections. Instead, I focus on recording and disseminating the wealth of information that the project generates.

The website has two purposes: it has a public face for the project, and private space for team members. The publicly accessible part contains information about our work and key research topics, abstracts from seminars and links to related sites. The site receives more hits than I anticipated, an average of fifteen thousand every month. I receive a variety of enquiries, some from scientists who want to know more about our work, some from school students doing projects. Occasionally someone who has found a fossil will write in wanting to know if it's a human specimen. The team members' web area contains a notice board, published and unpublished manuscripts, an archive of technical aspects of the project, minutes of meetings, posters we're producing, and electronic contour maps that I built using topographic information from the Ordnance Survey. People use these for research purposes and for creating illustrations.

The database is a GIS programme that condenses the information produced by the research project. I have mapped 253 localities so far. Each site is linked to the relevant records: where it is (the National Grid co-ordinates); the Mammal Assemblage-Zone (MAZ); the Marine Isotope Stage (where known); whether archaeology is present and what cultural type it is; any other information on human presence; a broad classification of the landscape – was it a river, lake, cave; and a list of the mammal fauna. For each mammalian species the order, taxon, family, common name and authority (where the information came from) is provided.

The database is work in progress. It will grow, as people submit their findings, to become an extremely useful research tool. It will be capable of selecting and mapping all the sites that fulfil certain criteria, so that we can search for meaningful patterns and correlations. For example, a researcher might ask for those sites that have bears, with human archaeology, in a cave, in a specific marine isotope stage. The relevant sites will appear plotted on either a simple outline map of Britain or a more complex contour map in closer focus. In 2007, the AHOB team will publish a major monograph of their work. We will compile and publish the database alongside, either online or on a CD which will accompany the monograph.

My experience of AHOB has been somewhat different from everyone else's. I work slightly outside the team but, at the same time, collaborate with everybody. It's been an excellent project to work on and the outcome – an organized and reliably dated mammalian and archaeological record of the British Quaternary – is incredibly important. This will give a much sounder basis for research into the effects of climate change and its impact on both humans and animals.

ROBERT SYMMONS

Robert Symmons, affectionately referred to as Nobs by his friends and colleagues, was AHOB's research assistant for the first four years of the Project. Although he variously describes himself as a dogsbody and dirt monkey, it's widely acknowledged that he has been an indispensable member of the team. Despite his comment at the end, he recently left us for an equally wonderful job as Curator of Archaeology at Fishbourne Roman Palace. He has been ably replaced by Silvia Bello and Mark Lewis.

I first became interested in archaeology at the age of six when my great-uncle Pete, himself a keen amateur archaeologist, showed me a small tin of Neolithic carbonized cereal seed. I had no idea that things could be that old and was instantly hooked. Aged fifteen I started working on excavations and after school spent my gap year working on digs in the UK and abroad. On one dig, in Jordan, the animal bone specialist Paul Croft got me interested in zooarchaeology. After that I went on to study for a degree, Masters and PhD at the Institute of Archaeology at the University of London. My speciality is taphonomy, the study of what happens to an animal's bones between its death and their recovery. I'm particularly intrigued by the way bones fragment and by what happens to the bones we never find.

Towards the end of my PhD, I had the most enormous stroke of luck. An email advertising for a research assistant to work on the AHOB project was circulated around the university. I applied and was successful. As research assistant, I try and help any member of the AHOB team in any way I can. The best thing about my job is the sheer variety. I might spend an entire day producing complex drawings at the computer or peering down a microscope, but the next day I'll be out in a field swinging a pick. There's no chance of getting bored.

One of my favourite jobs involves illustration. I produce maps, section drawings, site plans and diagrams for most of the team's publications. I also enjoy getting out in the field, although some of the digs I've been involved with in East Anglia have been unbelievably challenging. I've spent over fifteen years working as a dirt monkey and never has this been such a truer description. The defining feature of AHOB sites is that they tend to be astonishingly muddy, at times up

to your knees. I remember Simon Lewis trying to climb out of a trench at Hoxne using a ladder that just sank further into the mud with every step up. It took nearly half an hour to dig the ladder out. We likened it to the trenches of World War One. If Hoxne was the muddiest then Happisburgh was the coldest. The site is situated near the low tide mark, so one January I found myself standing in the North Sea, sieving mud, in bitter winds and horizontal rain whilst wearing a Russian military hat and inappropriate footwear. I thought I was going to die. But despite the privations I've been particularly excited about this aspect of the project. The very early sites on the East Anglia coast must rate as one of AHOB's major contributions, although of course the work has raised as many questions as it has answered.

When AHOB started I did a huge amount of sample processing. In fact for the first eighteen months, when people asked what I did, I was able to say, truthfully, that I picked mouse teeth out of buckets of sand. Samples are brought back from the field sites in big muddy sacks. I then put everything through a sieving machine. This is a big tank with a square sieve in the bottom that processes 10–15 kilos of sediment at a go. In the lid of the tank is a garden sprinkler to wash all the mud away, leaving sand and other particles.

I then dry what remains in the lab, wash it with fresh water, and dry it again. Next, I run the mixture through an 8mm sieve, then a 4mm sieve, then a 1mm and 0.5mm sieve. The rest, I throw away or take down to my allotment. (The allotment's now covered with sand from around the country. The vegetables don't seem to be experiencing any benefit, but it might pose an interesting conundrum for a geologist one day in the future.)

Once I've separated the different grades I pick out the interesting material, things such as animal bones, fish scales, insect remains, mollusc shells, seeds, chips of flint and little bits of the sieve. The smallest particles (0.5-1mm) are sorted with tweezers under a microscope, grain by grain. This is probably my least favourite job – it's extremely tedious and you often find very little. We've had some volunteers working on the project, all of whom have been exceptional, and I've roped them in to help me with the sample sorting whenever possible.

Other things I do include taking minutes at all the meetings, driving stuff around, and transcribing Roger Jacobi's data to computer. Roger doesn't use computers but writes everything out in beautifully neat longhand. His paper on Gough's cave is 31,408 words long and it took me weeks to type but it was worth it because it's such a brilliant piece of work.

I have the occasional moment, sitting in a meeting or listening to others' conversations, when I get a wonderful reality check. In all my born days I never thought I'd be involved in something so exciting and so important. This is my first job after my PhD and I think it's fair to say I'm not a dirt monkey anymore. It's the best job in the world.

SIMON LEWIS

Simon Lewis is a geologist who specializes in the Quaternary, and had already worked with many other potential members of the Project. His investigations of ancient environments provide a picture of the landscapes in which Britain's early humans lived.

I've always been fascinated by mountains and rivers and how they form and evolve over time. While studying for a geography degree this interest in geomorphology led me to focus on the Ice Ages because so many of the landscapes we see in Britain were created by glacial or cold climate processes. I decided to do a PhD so that I could continue research in this area, and I now work as a lecturer in physical geography at Queen Mary, University of London.

My expertise is in river sediments. I look at river deposits to see what they tell us about how rivers behaved and how key variables, such as flow and sedimentation,

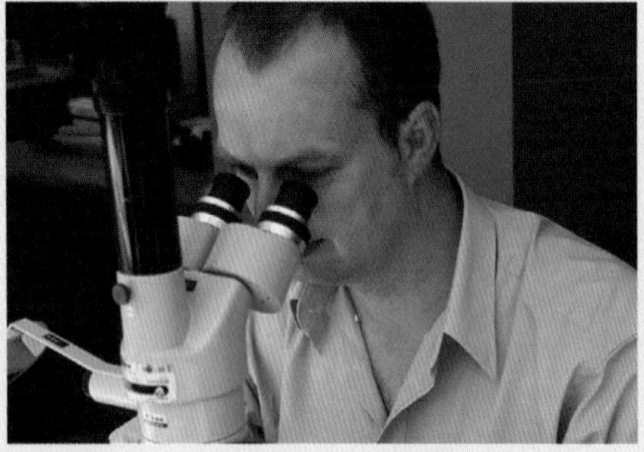

respond to changes in climatic and environmental conditions. Some of those river sediments contain archaeology and since completing my PhD I've also worked on a range of archaeological projects.

I've been involved with AHOB since its inception. My job is to interpret the sediments to determine a detailed stratigraphic sequence. A secure geological framework is essential as it provides a context for human occupation in terms of both environment and time.

My first project for AHOB was to conduct investigations at Hoxne, along with Nick Ashton and Simon Parfitt. There were unresolved questions about the age of sediments and how the archaeology fitted into the geological sequence. The excavations were awful because it was so muddy, but the work was extremely rewarding as it considerably improved our understanding of the stratigraphy. Hoxne and two roughly contemporaneous sites I worked on in Suffolk, Barnham and Elveden, are all located by rivers. It seems likely that during the Hoxnian interglacial (around 400,000 years ago) humans in this region preferred to live by rivers. Watercourses would have attracted prey animals, riverine erosion would have exposed useful raw materials like flint, and river channels would have made navigating the densely wooded landscape considerably easier.

One of the most exciting sites I have worked on is Norton Subcourse in Norfolk. This is a working sand and gravel quarry which revealed a river channel and a rather splendid exposure of sediments that, at 700,000 to 800,000 years old, relate to the very earliest part of the timespan covered by AHOB. Paradoxically, given the remit of our project, we haven't found any evidence of human occupation at this locality, but the site is extremely important because it provides information about the landscape and environment that existed when humans started living in the British Isles for the first time. It's a close neighbour of the Pakefield site, both in terms of geography and time. Now that the archaeology found at Pakefield has moved the earliest occupation back to around 700,000 years, this environmental information is critical. The sediment stack enabled me to build up an extremely detailed picture of the local landscape. Shallow marine gravels at the base are overlain by interglacial fluvial clays and silts, woody peat and organic silts. This is overlain by sands and gravels and finally glacial deposits from the ice age that followed.

The climate at the time the interglacial channel was active was warm, possibly slightly warmer than today. A section of river was cut off and abandoned, forming a body of still water. Reeds and alder grew on the muddy banks and hippos wallowed in the water. We've also found the remains of insects, fish, birds, reptiles, small mammals, horses and an elephant, and from their coprolites there's evidence of hyaenas.

I worked at Lynford, another Norfolk site that dates from the younger end of the project. It is located in a working quarry and is a Neanderthal site that dates to around 60,000 years ago, the early part of the last cold stage. This was when humans first returned to Britain after what seems to be a long and inexplicable absence during the preceding interglacial. The humans shared their landscape with cold-climate animals such as reindeer, horse, woolly mammoth and woolly rhino, and they left an abundance of artefacts. It's likely the Neanderthals hunted or at least scavenged the large mammals but as yet there's no firm evidence of butchery. The cold climate created a harsh landscape of sedge, grasses and weeds, with no trees or shelter, which raises interesting questions as to why people chose to live there. That's where my job ends and the archaeologists take over.

And that's why the AHOB project has been such a great success – because it's a team effort. By combining the work of experts in different fields we've gained a much more coherent understanding of the subject. We could all answer the project's questions individually, from our own perspectives, but by integrating our knowledge we've created a uniquely detailed picture.

MIKE RICHARDS

Isotope studies now have a key role in reconstructing the past – we saw in Chapter 2 how oxygen isotopes in deep-sea cores have helped to unravel the complexity of past climate changes. Mike Richards is one of the world's leading experts in the analysis of isotopes in bone to reconstruct ancient diets, and the work of him and his research team has formed an important part of AHOB.

I was born in Zambia and grew up in Wales and Canada. I started studying archaeology at Simon Fraser University in Vancouver, Canada, in 1988, and later moved to the UK to do a PhD in the Department of Archaeology at Oxford. I was a Lecturer, Reader and then Professor at the University of Bradford, but am now a Professor at both the University of Durham and the Max Planck Institute for Evolutionary Anthropology in Leipzig, Germany. I moved to Germany halfway through the AHOB project but I travel back to the UK frequently, so I don't feel left out at all.

I think that there are many people about my age who first got interested in archaeology the same way, inspired by the Tutankhamun exhibit that toured Europe and North America in the 1970s. My expertise is archaeological science or the application of scientific methods to archaeology, particularly the use of chemical analysis of bone to determine past diets in humans and animals. I got into this because I didn't want to limit myself to any particular region or period. Also, I really enjoyed the archaeological science courses I took at university, as it seemed to be a way to produce clear, unequivocal, facts about the past.

Chris Stringer invited me to join the AHOB project in 2002. I submitted a proposal for a study of stable isotopes that was approved by the other team members. Stable isotope analysis is one of the few methods that can give us direct measures of people's diets in the past. For the project I have mostly worked on carbon and nitrogen isotopes found in collagen, the organic content of bone. Like DNA, collagen degrades, so after about 100,000 years there isn't much left. But the little that remains has great potential for revealing what people and animals ate over their whole lifetimes, and in what general proportions. It's a relatively new area of research, and very exciting.

We also developed novel techniques for measuring oxygen isotopes in fossil bone and teeth, and are still applying these methods to AHOB material. This, combined with carbon and nitrogen analysis, can be used to determine past climates. The oxygen work was mainly undertaken by Vaughan Grimes, a Bradford PhD student funded by AHOB.

Most of the carbon and nitrogen in bone collagen comes from food the individual ate. The ratios of the stable isotopes of the two elements can be used as 'signatures' that provide a record of long-term diet. We can tell if the diet was mainly based on protein from the sea or land, and also whether the protein was mainly derived from animal or plant sources. Previous research, using these methods, had told us that European Neanderthals were carnivores at the very top of the food chain. Research I've done for AHOB on contemporaneous and more recent modern humans indicates that they had a similar diet, but also consumed aquatic resources, especially fish. There is no evidence indicating that Neanderthals ate any significant amounts of fish, so this may be a uniquely modern human adaptation in Palaeolithic Europe.

We also looked at two late Upper Palaeolithic British sites, Gough's Cave and Kendrick's Cave, that date to around 14,000 years ago. The people at Gough's ate mainly the meat of herbivores such as cattle and deer. There are lots of horse remains at the site but, surprisingly, the humans probably weren't eating them. The human bones at this site had cut marks and were deposited with butchered animal bones. So perhaps the people living at Gough's Cave did eat horse, but the human remains we've found belong to people who were killed and brought to the cave by its inhabitants.

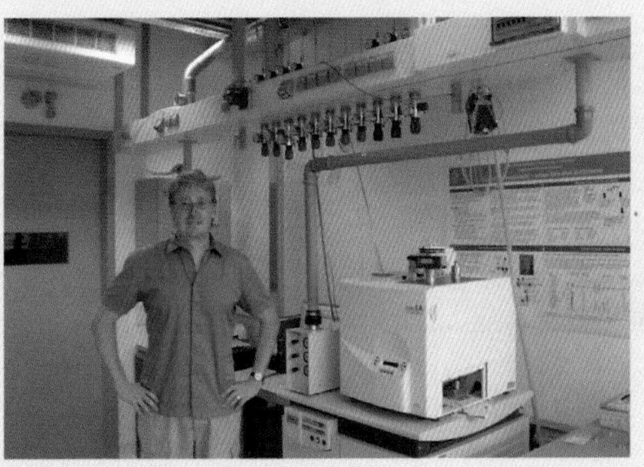

Our work on Kendrick's Cave showed that about 30 per cent of the food eaten by these people came from the sea. High nitrogen isotope values indicate that they ate marine mammals such as seals. This is fascinating and provides some of the earliest evidence for the significant use of marine mammals by humans.

AHOB's main achievement, from my perspective, has been the multidisciplinary nature of the project, which has brought together researchers from different fields for a common aim. It has helped us improve our methods of working on isotope analysis of very old material. In the future I would like to focus on extracting proteins other than collagen from bones and teeth that might survive longer than 100,000 years.

IAN CANDY

Ian Candy is another expert in stable isotope analysis, as well as in uranium-series dating and Quaternary sedimentology. To many of his colleagues, his methods may seem more like sorcery than plain old science. But by working his magic, Ian has been able to reconstruct Britain's past climates in extraordinary detail.

I first became interested in Quaternary geology while studying geography at Royal Holloway, University of London. After I graduated, I stayed on to work for Jim Rose, who later brought me onto the AHOB team. The ancient river deposits that I looked at in eastern England have subsequently become key sites for the AHOB project. I went on to do a PhD looking at terrestrial carbonates in southern Spain. Coincidentally, this work too has had a significant impact on AHOB research.

My work for AHOB has focused on investigating what stable isotopes can tell us about ice age rainfall and temperatures. The stable isotopes come from carbonates (limestone), which are minerals created by natural processes. The carbonates I've studied for AHOB come either from mollusc shells or are formed in soils, ground water, or in association with springs, called tufa.

Soil carbonates form as small nodules, about 2–3 centimetres long. The nodules are produced only under certain climatic conditions, so their presence tells us something about the climate at the time. As carbonates are extremely soluble, they don't accumulate in wet conditions, forming only where there is either low rainfall or very seasonal rainfall with long dry periods. In modern-day Europe they are restricted to some southern and eastern areas. In Britain, where rain falls throughout the year, all the carbonates in a soil are dissolved and washed away.

When I first found soil carbonate nodules in fossilized soils in Britain, I recognized them from my work in Spain. They immediately told me that the environment of the time had different rainfall patterns from today. That's why carbonates are so useful – insect and fauna studies can tell us about temperatures but they rarely reveal anything about precipitation. The analysis of soil, ground water and spring carbonates involves three processes. First, I log the carbonates in the field by drawing and describing how they occur in the sediments. This gives vital clues to the way in which they formed, that is within a soil profile or at the groundwater table. Next, I take the nodules to a laboratory and cut them into thin sections. I impregnate one section with resin, to make it hard. I then cut the section in half with a circular saw, polish the cut face, stick it to a piece of glass, and grind it down to a thin layer.

Under a microscope I can see structures within the nodule not visible to the naked eye. The shape and size of the crystals that form within a soil carbonate will be very different from those that form within a groundwater carbonate. Microscopic analysis, therefore, allows us to establish the nodule's origin and to identify evidence for alteration that may have occurred after the nodule formed.

The final stage is isotopic analysis. Isotopes are species of elements that have different atomic weights.

Because of this weight difference, natural processes selectively uptake the lighter isotopes. In soil carbonates, for example, the ratio between oxygen isotopes (^{16}O:^{18}O) is controlled by temperature, along with other variables, whilst the ratio between carbon isotopes (^{12}C:^{13}C) is controlled by the composition of the overlying vegetation. Isotopic analysis can, therefore, give us important information on the climatic and biological processes operating at the time the isotopes were formed.

I conducted an in-depth investigation at Pakefield and found that the soil nodules are particularly enriched in ^{18}O, indicating very warm temperatures and/or seasonal dryness. Pakefield is the earliest site of human occupation in all of northern Europe, and one of AHOB's most exciting ventures, so information about its climate is critical. We now have fantastically detailed information about both temperature and rainfall, thanks in part to the analysis of the soil nodules. In combination with the plant and animal evidence they showed that there was a Mediterranean-type climate, with a strongly seasonal precipitation regime – wet winters and dry summers. It is thought that humans lived in the Mediterranean regions of Europe for a long time before they migrated to the relatively cold north. Now we know that humans colonized northern Europe quite early, but at a time when the north enjoyed a warm climate that they were adapted to and would have found comfortable. This makes sense, and allows a new interpretation of the early appearance of humans in Britain.

I've also done a lot of work on the Hoxnian interglacial, an episode with a large number of occupation sites. I compared soil, ground water and tufa carbonates across the sites and found that the oxygen isotope ratios were not that dissimilar to the present day, although slightly heavier, possibly indicating a somewhat warmer climate.

Something else I hope to pursue in the near future is a dating strategy that involves earthworms. Dating is often a problem because of the lack of suitable material. Earthworms might be able to help: as they move through the ground, digesting soil, they precipitate carbonates as little granules, about 1–2 millimetres in size, which are excreted into the soil where large amounts accumulate. We're trying to develop a technique to date these deposits using uranium-series dating. If that were possible it would offer an amazing opportunity to accurately date sites of human occupation, because of the ubiquity of earthworms. At the moment, we have no idea how well it might work, but I'm keen to find out.

Working on AHOB, with people who know the sites and the material so well, has been incredibly exciting. Quite often the archaeologists or mammal specialists will be sieving material from their sites and they'll alert me to material I wouldn't have known about otherwise. It's wonderful to have access to material from so many important sites.

My priority now is to develop new techniques with which to date and reconstruct the climates of a range of sites. AHOB has provided the opportunity to start research that, in the future, could lead to some really big advances.

RUSSELL COOPE

As well as the core members of the Project we have met above, we also have sixteen associate members, who are our closest collaborators. I've chosen just one of these to show how important they are, too. As perhaps the world's leading Quaternary palaeoentomologists, there are few people who know more about fossil insects than Russell Coope. From a rambling farmhouse in the remote Scottish Highlands, Russell makes detailed reconstructions of Britain's ancient environments and climates by studying these fascinating animals.

I have been studying Quaternary insect fossils for over fifty years. I started work on them almost entirely by accident. I was washing mud off mammoth and woolly-rhino bones, when I discovered that it contained large numbers of beautifully preserved insects, mostly beetles. Very few people had done any systematic work on these fossils and I was immediately hooked.

The mud that most archaeologists shovel up and barrow off the site I take home because it contains real treasures. Beetles tend to be large and robust, rather like entomological tanks, so they make excellent fossils and objects for study. I started off assuming that because the beetles were associated with extinct animals, they would also be extinct. Some time later, I realized that almost all of them were still living today. During the Quaternary their rate of evolution seems to have been incredibly slow; some of today's species were living more than a million years ago and, in the beetle world, one million years represents one million generations. So there was plenty of opportunity for evolution, but for some reason it didn't happen, at least in the features we can observe in the fossils.

To identify fossil beetles I compare them with well-authenticated modern specimens. Their complexity means that I can often identify them to the species level. The most exciting fossils, though, are those which are highly distinctive but do not match any familiar species. When that happens, I widen the search to include distant international collections and literature. One memorable example is a highly unusual dung beetle that was abundant in the British Isles during the middle of the last Ice Age. The identity of this beautiful animal, which sports a large distinctive horn, split into two, in the middle of its head, remained a puzzle for more than fifteen years. As it often does, the solution came by accident. One day, while idly browsing the beetle collections in the Natural History Museum, I found four identical specimens. The label on the pins read, 'Dingri, 10,000 feet, 1925 British Everest Expedition'. To find that a Tibetan dung beetle once inhabited Ice Age Britain was indeed a surprise.

My growing involvement with Quaternary beetles led me to realize how useful they are as indicators of past environments. Beetle species are extremely fastidious about where they live and what they do for a living. So if the fossil species have the same environmental requirements as their modern descendants, I can reconstruct the environment they lived in.

We know that ancient beetles required the same environmental conditions as their modern counterparts because of the way assemblages have remained constant. Whole suites of beetles from the British Ice Age still live in association with each other, often thousands of miles from where the fossils are found – some are now found in the high Arctic and some on the Siberian tundra. As temperatures changed, each species moved separately, tracking the acceptable climate across continents. After their forced marches the species reassembled in familiar associations in their new locations. An evolutionary explanation for this phenomenon is so unlikely that it can be discarded.

The great significance of this is that fossil beetles can give precise indications of ancient climates. The majority of species are 'hard wired' with strict thermal requirements, so their presence allows an accurate reading of past temperatures. To take that reading, I use the Mutual Climatic Range method. This plots the present-day range of each species on climate space whose limits are defined by temperature gradients rather than latitude and longitude. In this way, geographically disparate species that live in similar climates, for example animals from cold northerly latitudes and those that live at altitude further south, can overlap in the same area of climate space. By investigating this overlap, I can calculate historical temperatures with surprising accuracy. Mean July temperatures can often be estimated to within two or three degrees. Winter temperatures are often more difficult and less precise, perhaps because the animals hibernate and are less affected by their thermal environment during these months.

As well as climate, fossil beetles tell us about their local habitat. There are some water beetles that are entirely dependent on running streams, and others which need stationary water. Some water beetles are carnivorous, while others eat decomposing vegetation. Their terrestrial counterparts include carnivores, reed-eaters, tree-eaters, dung feeders, beetles which eat only fresh carcasses, beetles which eat only dry carcasses, and others which eat only the maggots which eat the carcasses, and so on down the food chain. Together, the beetles enable me to create a mosaic picture of their ancient environment.

I became involved with AHOB because many of my insect faunas are associated with evidence of human occupation, so the palaeoenvironmental evidence they provide has significance for understanding past living conditions. Over the years I naturally found myself working with various members of the AHOB team. They have proved most valuable contacts and enjoyable company. The project itself has been great fun. As a corporate effort, it is possible to analyse ancient history in much more depth, and it's always more exciting to look at the whole picture, not just the minutiae. The beetle assemblages are important because they can often yield environmental information that cannot be deduced from other fossil evidence.

The Lynford site provides an excellent example of the unique contribution that the study of beetle assemblages can make. Enormous numbers of dung and carcass beetles were found here in association with the bones of large herbivorous mammals, chiefly mammoths, alongside numerous flint tools and flakes. There can be little doubt that both Neanderthals and mammoths were occupying this site at the same time. The beetles testify to the presence of vast quantities of dung, and indicate that the site was occupied during the summer months. The temperatures were comparable to present-day Siberia so life cannot have been easy. The beetles also show that after this period the climate became savagely cold with winter temperatures dropping below -20°C, but it is not known if humans were present during this period.

The beetle assemblage from Pakefield, the oldest known site of human occupation in Britain, is also very informative. I found both running-water and stationary-water beetles, indicating a river characterized by rapids and pools, meandering across a flood plain. There's also a highly specialized beetle that lives only on the submerged, decomposing trunks of oak trees, suggesting that the riverbanks were lined by large trees. Carcass- and maggot-feeding beetles suggest that large mammals were also present. Despite the site being on the coast today, there is no evidence of any salt marsh species, suggesting the sea was not that close 700,000 years ago. Interestingly, some of the species from Pakefield are now exclusively southern European. The Mutual Climatic Range indicates mean July temperatures of 17–23°C. The climate was several degrees warmer than the present day and more Mediterranean in character.

Beetles have also shown us that climates can change extremely rapidly, even without human intervention. About 15,000 years ago, as the last glaciation was drawing to a close, Britain's climate warmed dramatically. Mean July temperatures rose by 7°C in such a short time that I can't measure it precisely, but the whole rise probably occurred during the span of one human lifetime.

Finally, I must admit that one of the driving forces behind my work is the fact that beetles are stunningly beautiful animals. Even fossils dating from the start of the Pleistocene retain their vivid colours under the microscope. Species move on a vast scale, so there is the excitement of the hunt for them in far-flung exotic locations. From the scientific point of view, the main thrill of working with them is that, since they represent species still living, I can ask them meaningful questions and, what is more, receive useful answers back. You can't gain such precise information from more spectacular extinct animals, such as the woolly rhino or mammoth, because their environmental requirements can only be inferred. On the other hand, the reluctance of beetles to evolve or to become globally extinct means that they are less valuable as stratigraphical indices than are the rapidly evolving mammals.

There is a well-known story in entomological circles that when the famous biologist J.B.S. Haldane was asked what he thought of God, he replied, 'He must have an inordinate love of beetles.' Haldane was referring to the sheer number of species, but we now know that there is so much more to them than mere diversity.

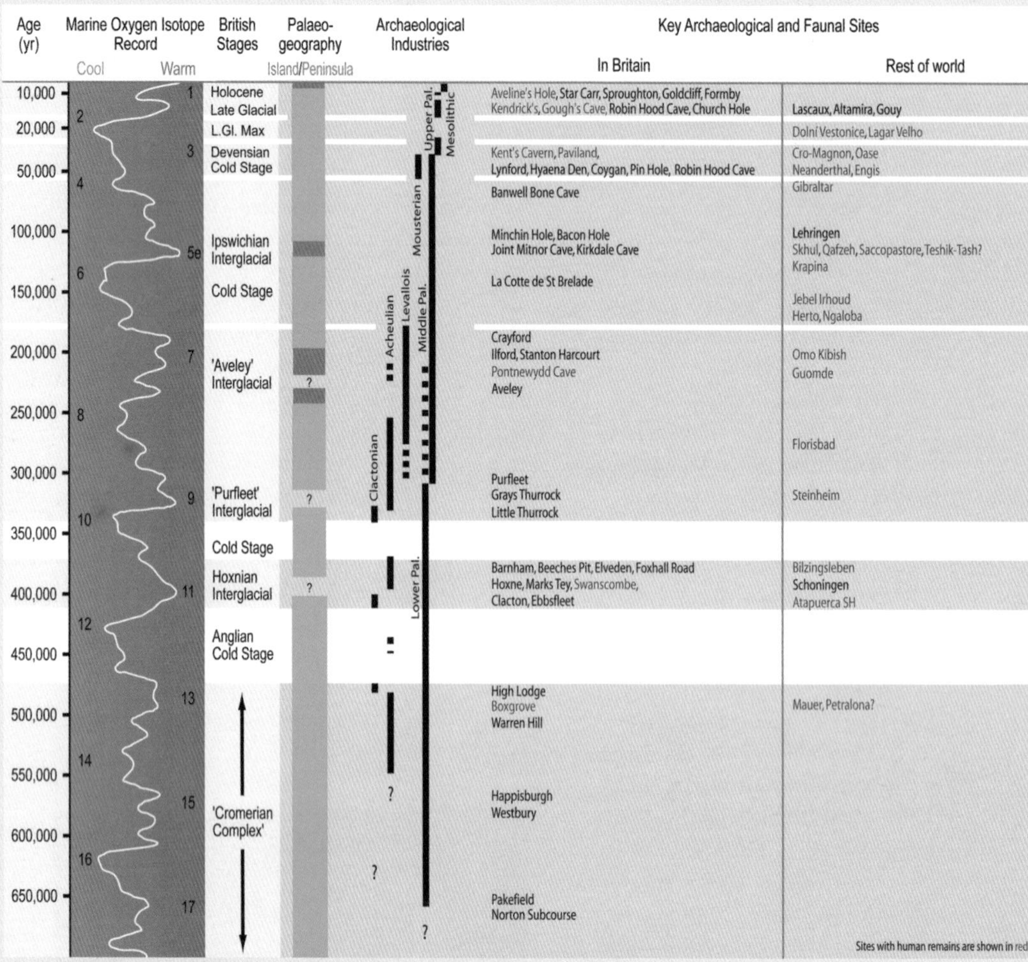

Age (yr)
Marine Oxygen Isotope Record
Cool
Warm
British Stages
Palaeo-geography
Island/Peninsula
Archaeological Industries
Key Archaeological and Faunal Sites
In Britain
Rest of world
10,000
20,000
50,000
100,000
150,000
200,000
250,000
300,000
350,000
400,000
450,000
500,000
550,000
600,000
650,000
1
2
3
4
5e
6
7
8
9
10
11
12
13
14
15
16
17
Holocene
Late Glacial
L.Gl. Max
Devensian Cold Stage
Ipswichian Interglacial
Cold Stage
'Aveley' Interglacial
'Purfleet' Interglacial
Cold Stage
Hoxnian Interglacial
Anglian Cold Stage
'Cromerian Complex'
?
Clactonian
Acheulian
Levallois
Mousterian
Upper Pal.
Mesolithic
Middle Pal.
Lower Pal.
Aveline's Hole, Star Carr, Sproughton, Goldcliff, Formby
Kendrick's, Gough's Cave, Robin Hood Cave, Church Hole
Kent's Cavern, Paviland,
Lynford, Hyaena Den, Coygan, Pin Hole, Robin Hood Cave
Banwell Bone Cave
Minchin Hole, Bacon Hole
Joint Mitnor Cave, Kirkdale Cave
La Cotte de St Brelade
Crayford
Ilford, Stanton Harcourt
Pontnewydd Cave
Aveley
Purfleet
Grays Thurrock
Little Thurrock
Barnham, Beeches Pit, Elveden, Foxhall Road
Hoxne, Marks Tey, Swanscombe,
Clacton, Ebbsfleet
High Lodge
Boxgrove
Warren Hill
Happisburgh
Westbury
Pakefield
Norton Subcourse
Lascaux, Altamira, Gouy
Dolní Vestonice, Lagar Velho
Cro-Magnon, Oase
Neanderthal, Engis
Gibraltar
Lehringen
Skhul, Qafzeh, Saccopastore, Teshik-Tash?
Krapina
Jebel Irhoud
Herto, Ngaloba
Omo Kibish
Guomde
Florisbad
Steinheim
Bilzingsleben
Schoningen
Atapuerca SH
Mauer, Petralona?
Sites with human remains are shown in red

ACKNOWLEDGEMENTS

First I would like to acknowledge the whole field of Quaternary Research, especially the Members and Associate Members of the Ancient Human Occupation of Britain project, and the Leverhulme Trust, without whom there would have been no project, and little for me to write about! A number of these colleagues have given me additional help with the book in the form of information, checking text and providing illustrations. In particular, Silvia Bello prepared a number of the illustrations, including the maps. I would especially like to thank Sarah Lazarus for her considerable time and trouble in interviewing us and compiling the personal profiles that make up most of the Appendix. I am also very grateful to the Natural History Museum and the many staff who have supported AHOB and me over the last five years, and this includes the NHM Photographic Unit for all their work on specimens and in the field and Richard Fortey for reading parts of the manuscript. Finally I would like to thank Angela Marshall and Mark Lewis for help with the proofs, John Brockman for his encouragement, Smith & Gilmour for design and the staff of Penguin Books, especially my editor Will Goodlad, for all their work in bringing this book to fruition.

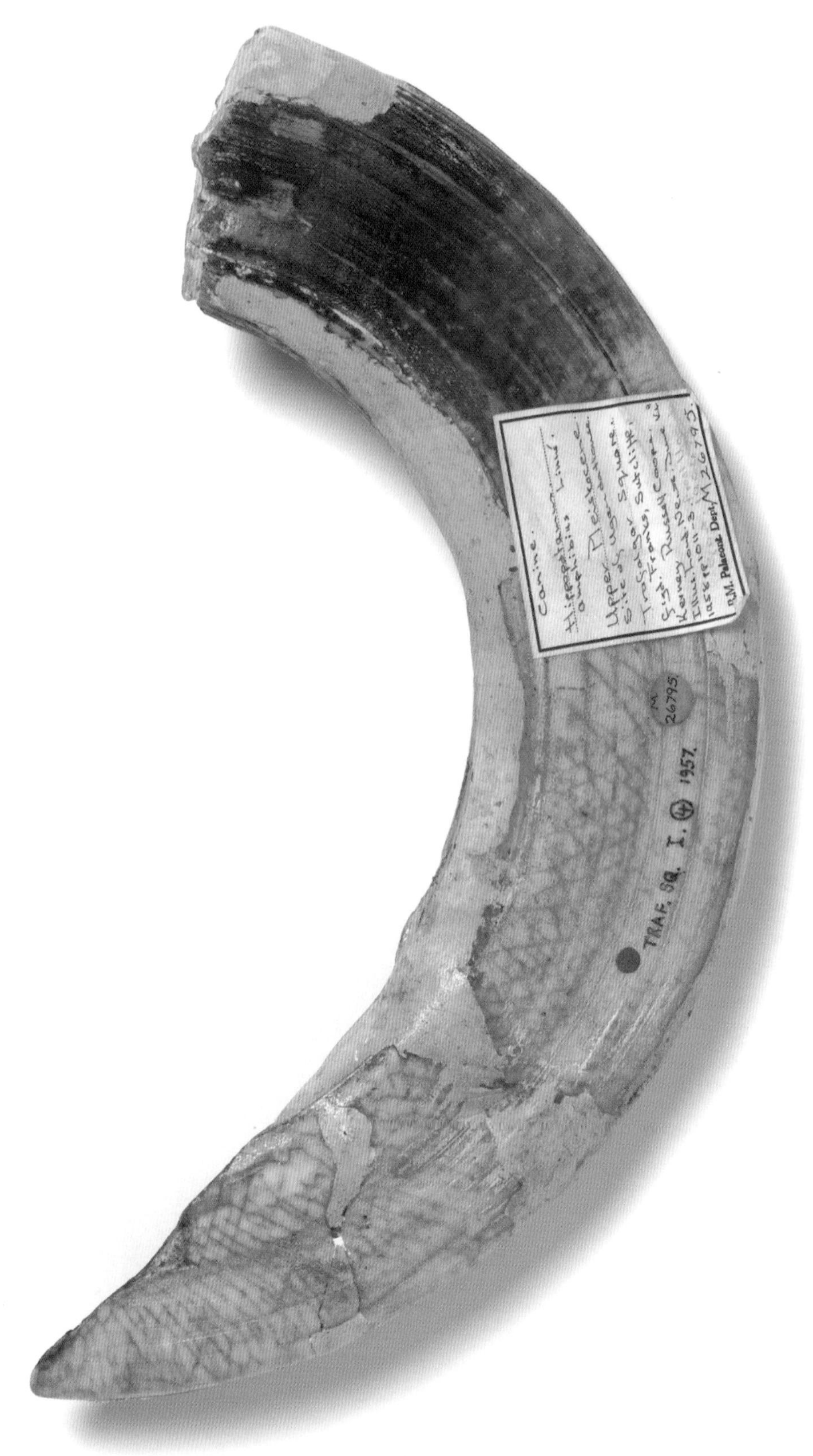
Canine.
Upper Pleistocene.
Trafalgar Square.
B.M. Palaeont. Dept. M 26795.
M. 26795.
TRAF. SQ. I. (4) 1957.

ILLUSTRATION ACKNOWLEDGEMENTS

Every effort has been made to contact all copyright holders. The publishers will be happy to make good in future editions any errors or omissions brought to their attention.

INTRODUCTION IN THE BEGINNING

12/13 *The Deluge*, Francis Danby, first exhibited 1840 (© Tate, London 2006)
16 Hand axes from Hoxne, Suffolk, 1800 (Wellcome Trust Medical Photographic Library)
18 Georges Cuvier, 1769–1832 (© Natural History Museum, London)
19 Georges Cuvier, lecturing (© Natural History Museum, London)
20 Portrait of William Buckland, Ansdell, *c.*1843 (© Geological Society/NHMPL)
21 (*top*) Kirkdale Cave, plate engraving from *Reliquiae Diluvianae*, Rev. William Buckland, John Murray, 1823 (© York Museums Trust, Yorkshire Museum)
21 (*bottom*) Buckland enters Kirkdale Cave (© Oxford Museum of Natural History)
22 Sir William Buckland and family, unnamed artist, in *Allers Family Journal*, 27 December 1927 (© Mary Evans Picture Library)
24/25 Diagrammatic section of the Earth's crust, fold-out plate from *Geology and Mineralogy Considered with Reference to Natural Theology*, Rev.William Buckland, 1836 (© Natural History Museum, London)
26 Sir Charles Lyell, 1797–1875 (© Geological Society/NHMPL)
27 William Pengelly, 1812–1894 (© Natural History Museum, London)
28 John Philp's promotional poster for Brixham Cavern (© Natural History Museum, London)
30 Map (Silvia Bello © AHOB)
31 'Awful Changes', cartoon by H. T. de la Beche, frontispiece, *Curiosities of Natural History*, Francis Buckland (© Natural History Museum, London)
33 Charles Darwin (© Wellcome Trust Medical Photographic Library)
34 Map of Britain from *Abbreviatio chronicorum Angliae* by Matthew Paris, Monk of St Albans, 1250–1259 (© The British Library)
35 'Delineation of Strata of England and Wales', William Smith, 1815 (© Natural History Museum, London)
38 Alpine sequence of glacials, 1964 (© Natural History Museum, London)
40 Neander Valley postcard (CBS)
43 Ballochmyle Quarry, Mauchline, Ayrshire, Permian red desert
sandstones showing large-scale aeolian (wind-blown) dune-bedding, 1921 (© Royal Geological Society)
44 Mauer jaw (© Natural History Museum, London)
46 'Searching for the Piltdown Man' (© Natural History Museum, London)
47 Piltdown man cranium and mandible as reconstructed by Dr. A Smith Woodward and by Professor Arthur Keith (© Natural History Museum, London)
48 *A Discussion of the Piltdown Skull* (© Natural History Museum, London)
50 Geologists' Association visit to Piltdown, 12 July 1913 (© Natural History Museum, London)
51 Piltdown Man public house, Uckfield, East Sussex (© Natural History Museum, London)

CHAPTER ONE
THE FIRST BRITONS

52/53 River meanders through trees (© Getty Images)
56 Dmanisi, Georgia (by courtesy of Professor David Lordkipanidze)
57 Map (Silvia Bello © AHOB)
60 Casts of Gran Dolina skull fragment and jaw (© Natural History Museum, London)
62 Happisbugh, Norfolk, March 2006 (© Mike Page)
63 Black flint handaxe, Happisburgh (Benoît Audureau © Natural History Museum, London)
64 Black flint handaxe back, Happisburgh (Benoît Audureau © Natural History Museum, London)
65 Happisburgh beach (© CBS)
66 Bison bone (Benoît Audureau © Natural History Museum, London)
67 Map (Silvia Bello © AHOB)
68/69 Hippos among dense delta vegetation in Botswana (Getty Images)
70 Hyaena jaw (Benoît Audureau © Natural History Museum, London)
71 Pakefield tools (Benoît Audureau © Natural History Museum, London)
72 Beetle specimens (© AHOB)
74 Rhino jaw (Benoît Audureau © Natural History Museum, London)
75 *Nature*, 15 December 2005 (© *Nature*)
76 Species migration diagram (© M. Lahr/R. Foley)

CHAPTER TWO
UNDERSTANDING ICE AGES

78/79 Glacier, Antarctica (© Getty Images)
82 Milankovich diagrams, Geoff Penna (© Thames & Hudson Ltd, London. From *The Complete World of Human Evolution* by Chris Stringer and Peter Andrews, Thames & Hudson, London and New York)
84 The *Glomar Challenger* (© Bettman/CORBIS)
86 Polarized light micrograph of an ice sample extracted in Antarctica, 1978 (© CSIRO/Science Photo Library)
88 Map (Silvia Bello © AHOB)
89 Vole bones (© AHOB)
91 Author at Westbury, 1971 (© Natural History Museum, London)
92/93 Herds of buffalo crossing river, Okavango Delta, Botswana (© Getty Images)
94 Boxgrove scatter (© Boxgrove Project)
95 White flint stone tool (front), Boxgrove (Benoît Audureau © Natural History Museum, London)
96 (*top*) White flint stone tool (back), Boxgrove (Benoît Audureau © Natural History Museum, London)
96 (*bottom*) *British Archaelogy*, October 1996 (© *British Archaeology*)
97 Horse bone with cut marks (© Boxgrove Project)
98 (*top*) Red deer antler hammer (Benoît Audureau © Natural History Museum, London)
98 (*bottom*) Rhino pelvis (Benoît Audureau © Natural History Museum, London)
99 'Roger', Boxgrove human tibia (Benoît Audureau © Natural History Museum, London)
100 (*top*) Human teeth from Boxgrove (© Natural History Museum, London)
100 (*bottom*) Boxgrove Man cartoon (*Evening Standard*)
101 Boxgrove excavations (© Boxgrove Project)

CHAPTER THREE
THE GREAT INTERGLACIAL

102/103 A line of African elephants march through grasslands (© Getty Images)
106 Aerial view River Thames, London (© GeoPerspectives 2006)
108 (*left*) Alvin Marston (© English Nature)
108 (*right*) Alvin Marston and his son John (© Nature Conservancy)
109 (*top*) Swanscombe flint handaxe (Benoît Audureau © Natural History Museum, London)
109 (*bottom*) Swanscombe (© Natural History Museum, London)
110/111 Swanscombe skull bones (Benoît Audureau © Natural History Museum, London)
113 Descending the ladder in Atapuerca cave (© Atapuerca Project)
114 Skull casts (© Natural History Museum, London)

116 Juan Luis Arsuaga and Stephen Aldhouse-Green, Atapuerca (CBS)
117 Atapuerca finds (© Javier Trueba, Science Photo Library)
118 Cast of Skull 5 Atapuerca (© Natural History Museum, London)
120/121 A herd of lechwe running through a flooded plain (© National Geographic)
122 Map (Silvia Bello © AHOB)
124 Tools (© Natural History Museum, London)
125 Clacton spear (Benoît Audureau © Natural History Museum, London)
126 Dr. Hartmut Thieme with Schöningen spear (Hartmut Thieme)
127 Map (Silvia Bello © AHOB)
128 Levels of rock, Swanscombe (© AHOB)
129 Mollusc (© AHOB)
131 Three Swanscombe handaxes (front) (Benoît Audureau © Natural History Museum, London)
132 Three Swanscombe handaxes (back) (Benoît Audureau © Natural History Museum, London)
134 A prepared core and a detached flake (Benoît Audureau © Natural History Museum, London)
135 (*left*) Prepared core (Benoît Audureau © Natural History Museum, London)
135 (*centre*) Neanderthal handaxe (Benoît Audureau © Natural History Museum, London)
135 (*right*) Swanscombe handaxe (Benoît Audureau © Natural History Museum, London)

CHAPTER FOUR DESERTED BRITAIN

136/137 Kamchatka Peninsula, Russia (© Getty Images)
142 Rhino humerus and elephant vertebra (Benoît Audureau © Natural History Museum, London)
143 Aurochs horn and skull (Benoît Audureau © Natural History Museum, London)
145 Map (Silvia Bello © AHOB)
147 Pontnewydd cave, plan (© Pontnewydd Project)
150/151 Migrating through the snow, Nunavut, Canada (© Getty Images)
152 Pontnewydd teeth (© Natural History Museum, London)
154 Mammoth jaw (Benoît Audureau © Natural History Museum, London)
156 Rhino tooth with flint tools (Benoît Audureau © Natural History Museum, London)
157 Ouaisné, Channel Islands (© *Jersey Evening Post*)
158 (*left*) the cliffs at La Cotte de St Brelade (© Société Jersiaise)
158 (*right*) La Cotte excavation team (© Société Jersiaise)
159 La Cotte teeth (© Natural History Museum, London)
161 Strait of Dover and English Channel (© NASA/Science Photo Library)
162 Channel survey diagram (© Jean Francois Bourillet)
164 Reverend Wiliam Buckland (© Natural History Museum, London)
165 (*top*) Towards Victoria Cave (© Brian R. Hill)
165 (*bottom*) Hippo tooth, Trafalgar Square (front) (Benoît Audureau © Natural History Museum, London)
166 (*top*) Prehistoric Trafalgar Square, *Illustrated London News* (© Natural History Museum, London)
166 (*bottom*) Hippo tooth, Trafalgar Square (back) (Benoît Audureau © Natural History Museum, London)
167 (*left*) Brown bear humerus, Germany (Benoît Audureau © Natural History Museum, London)
167 (*right*) Giant Bear humerus (part), Banwell (Benoît Audureau © Natural History Museum, London)
168 William Beard (© Banwell Caves Heritage Group)
169 Banwell bone pile (CBS)

CHAPTER FIVE NEANDERTHALS AND US

170/171 Briar Island, Nova Scotia, North America
174 Neanderthal and Cro-Magnon skulls (Benoît Audureau © Natural History Museum, London)
175 Map (Silvia Bello © AHOB)
176 Beetle collection (© Natural History Museum, London)
178 Mammoth tusk, Lynford (© Nigel Larkin/Phil Rye)
179 Bear tooth, Lynford (© Nigel Larkin/Phil Rye)
180/181 Siberian mammoth in St Petersburg Museum of Zoology (© Getty Images)

182 Black flint handaxe in ground, Lynford (© Nigel Larkin/Phil Rye)
183 Lynford strata deposits (© Nigel Larkin/Phil Rye)
184 Neanderthal handaxes (© Natural History Museum, London)
185 Goat Hole Cave, Paviland (© Wellcome Trust Medical Photographic Library)
186/187 Cave art, Chauvet, France (© Dr Jean Clottes)
188 Cast of Ivory figurine, Avdeevo, Ukraine (Benoît Audureau © Natural History Museum, London)
189 (*top*) Cast of Venus statue (front) (Benoît Audureau © Natural History Museum)
189 (*bottom*) Cast of Mammoth engraving (Benoît Audureau © Natural History Museum, London)
190 Cast of Venus statue (back) (Benoît Audureau © Natural History Museum, London)
191 Map (Silvia Bello © AHOB)
193 Neandethal skullcap (Benoît Audureau © Natural History Museum, London)
194 Skull comparison (CBS/Musée de l'Homme)
195 *Homo* comparison (© Natural History Museum, London)
196 Neanderthal Man (© *Daily Express*)
197 Stan Laurel, 1927 (© Getty Images)

CHAPTER SIX WHAT THEY GORGED IN CHEDDAR

200/201 Cairngorms National Park, Scotland (© Roger Antrobus/CORBIS)
204 *A View of Cheddar Gorge* (oil on canvas), George Vincent, 1796–1831 (© Yale Center for British Art, Paul Mellon Collection, USA/Bridgeman Art Library)
205 Richard Gough at the entrance to the caves complex *c.* 1900, Stanley Chapman (by courtesy of Dave Irwin)
206 Ladder at Gough's Cave (© Getty Images)
207 (*top*) Arthur and William with skull, photograph by Henry Collard (by courtesy of Dave Irwin)
207 (*bottom*) Reindeer antler bâton (Benoît Audureau © Natural History Museum, London)
208 Worked bone and antler, Gough's Cave (© Natural History Museum, London)
209 Bone needle (Benoît Audureau © Natural History Museum, London)
210 Cheddar Gorge (© tonyhowell.co.uk)
212 Human skull with cut marks (© Natural History Museum, London)
213 Newspaper coverage of Cheddar Gorge discoveries (© *Daily Telegraph*)
214 'Remarkable Discovery in Cheddar Caves' (© Natural History Museum, London)
216/217 Wild horses, Wyoming, USA (© CORBIS)
218 Cast of Bone spear-thrower (Benoît Audureau © Natural History Museum, London)
219 Map (Silvia Bello © AHOB)
220 (*left*) Creswell Crags cave art (© English Heritage and Creswell Heritage Trust, by courtesy of Paul Pettitt and Paul Bahn)
220 (*right*) View from Robin Hood Cave (© Creswell Crags Museum, www.creswell-crags.org.uk)
222 Tools, Gough's and Kent Cavern (© Natural History Museum, London)
224 Barbed points, Star Carr (Benoît Audureau © Natural History Museum, London)
225 Amber from Star Carr (© Natural History Museum, London)
226 Human jaw bone, Gough's Cave (Benoît Audureau © Natural History Museum, London)
228 Cheddar Man (Benoît Audureau © Natural History Museum, London)
231 DNA chart (© Martin Richards)
233 Formby footprints, trail (© Gordon Roberts)
234 Formby footprint, adolescent or young female (© Gordon Roberts)
236/237 Formby prints, animals and children (© Gordon Roberts)

CHAPTER SEVEN OUR CHALLENGING CLIMATES

238/239 Flooded fields, Cheshire (© Getty Images)
242 Europa, moon of Jupiter (© CORBIS)
244 Erik the Red (© Bettmann/CORBIS)
246/247 *A Frost Fair on the Thames at Temple Stairs*, Abraham Hondius, 1684 (© Museum of London)

248 (*left and right*) Arctic sea ice (© NASA/Science Photo Library)
250 'Drunken Forests', Siberia (© Rolf Hicker)
251 Polar bear (© ecoscene)
253 Thames Barrier, London (© Skyscan.co.uk)
254 Hurricane Alberto (© GEOEYE/Science Photo Library)
256 Heinrich Events graph (by courtesy of wikipedia)
259 Aletsch Glacier, Switzerland (© Getty Images)
261 Map (Silvia Bello © AHOB)
262 Mount Kilimanjaro (© Kazuyoshi Nomachi/CORBIS)
263 (*left*) Capercaillie (© Eric Dragesco/Ardea.com)
263 (*right*) Snow bunting (© Jim Zipp/Ardea London Ltd)
266/267 Satellite monitoring ozone above the South Pole (© NASA/Science Photo Library)
269 Keli Mutu volcano, Flores (© Tiziana and Gianni Baldizzone/CORBIS)
272/273 Brighton Beach, 19 July 2006 (© Kent News and Picture Agency)

APPENDIX
THE AHOB TEAM

274 Lynford strata (© Nigel Larkin/Phil Rye)
276 Chris Stringer (© Natural History Museum, London)
278 Andy Currant (© AHOB)
279 Andy Currant (© Natural History Museum, London)
280 Gough's Cave (© Natural History Museum, London)
281 Roger Jacobi (*right*) (© AHOB)
282 Simon Parfitt (© AHOB)
283 Jim Rose (© AHOB)
284 Digging at Happisburgh (© AHOB)
285 Danielle Schreve (© AHOB)
287 Nick Ashton (*right*) (© AHOB)
289 Mark White (© AHOB)
291 David Polly (© AHOB)
292 Robert Symons (© AHOB)
293 Robert Symons (© AHOB)
294 Simon Lewis (© AHOB)
295 Mike Richards (© AHOB)
296 Norton Subcourse (© AHOB)
299 Russell Coope (© AHOB)

ACKNOWLEDGEMENTS

300 AHOB chart (© AHOB)

ILLUSTRATION ACKNOWLEDGEMENTS

302 Hippo tooth, Trafalgar Square (front) (Benoît Audureau © Natural History Museum, London)

SOURCES AND FURTHER READING

308 *On the Origin of Species* manuscript, Charles Darwin (© Natural History Museum, London)

p. 259 See 5th edn 1869. 261

As to bodily organs. Changes of instinct may sometimes be facilitated by the same species having different instincts at different period of life, or time of the year, or when placed under different circumstances &c, in which case either the one or the other instinct might be preserved by natural selection: and such cases of diversity of instinct in the same species can be shown to occur in nature.]

[Again, as in the case of corporeal structure, & conformably with our theory, the instinct of each species is good for itself, but has never, as far as we can judge, been produced for the exclusive good of others.

(A) each species tries to take advantage of the instincts of others. In certain cases some instincts cannot be considered as absolutely perfect; but as details on this and other heads are not indispensable they shall not be here given.

As some degree of variation in instincts under a state of nature, & the inheritance of such variations, are indispensable foundation for natural selection, it would be highly desirable here to

p. 259 near bottom

p. 260

SOURCES AND FURTHER READING

GENERAL READING

N. Barton, *Ice Age Britain* (Batsford, London, 2005)

E. Delson, I. Tattersall, J. Van Couvering and A. Brooks (eds.) *Encyclopedia of Human Evolution and Prehistory* (second edn; Garland Press, New York, 2000)

D. Johanson and B. Edgar, *From Lucy to Language* (Weidenfeld and Nicolson, London, 1996)

R. Jones and D. Keen, *Pleistocene Environments in the British Isles* (Chapman and Hall, London, 1993)

R. G. Klein, *The Human Career* (University of Chicago Press, Chicago, 1999)

R. Lewin, *Human Evolution: An Illustrated Guide* (fourth edn; Blackwell Scientific Publications, Oxford, 1999)

S. Lewis and N. Ashton (eds.), *The Palaeolithic Occupation of Europe: In Memory of John J. Wymer 1928–2006,* special issue of *Journal of Quaternary Science,* 21 (2006; now available on-line at the Wiley website: http://www3.interscience.wiley.com/cgi-bin/jhome/2507). This volume, in memory of Associate Member John Wymer, contains nine contributions by AHOB Members or Associates on a number of different sites.

C. Stringer, *Hominid Remains – An Update,* Volume 3: *British Isles and East Germany,* Université Libre de Bruxelles, 1990, pp. 1–40

C. Stringer, 'Modern Human Origins: Progress and Prospects', *Philosophical Transactions of the Royal Society, London (B)* 357 (2002), 563–79

C. Stringer, 'The Ancient Human Occupation of Britain (AHOB) Project', *Transactions of the Leicester Literary and Philosophical Society* 99 (2005), 29–32

C. Stringer and P. Andrews, *The Complete World of Human Evolution* (Thames and Hudson, London, 2005)

C. Stringer and C. Gamble, *In Search of the Neanderthals* (Thames and Hudson, London, 1993)

A. J. Stuart, *Pleistocene Vertebrates in the British Isles* (Longman, London, 1982)

A. J. Sutcliffe, *On the Track of Ice Age Mammals* (British Museum (Natural History), London, 1985)

K. Willis, K. Bennett and D. Walker (eds.), *The Evolutionary Legacy of the Ice Ages* (Royal Society, London, 2004)

The AHOB website:
http://www.nhm.ac.uk/hosted_sites/ahob/

The National Ice Age Network website:
http://www.iceage.bham.ac.uk/home.html

INTRODUCTION IN THE BEGINNING

G. Daniel, *A Short History of Archaeology* (Thames and Hudson, London, 1981)

D. A. McFarlane and J. Lundberg, 'The 19th-century Excavation of Kent's Cavern, England', *Journal of Cave and Karst Studies* 67 (2005), 39–47

K. Oakley, 'The Problem of Man's Antiquity: An Historical Survey', *Bulletin of the British Museum of Natural History (Geology)* 9 (1964), 85–155

M. Pitts and M. Roberts, *Fairweather Eden* (Century, London, 1997)

J. Weiner and C. Stringer, *The Piltdown Forgery* (50th anniversary edition) (Oxford University Press, Oxford, 2003)

CHAPTER ONE THE FIRST BRITONS

(various authors) 'Pakefield: A Weekend to Remember', *British Archaeology* 86 (2006), 18–27

S. C. Antón and C. C. Swisher III, 'Early Dispersals of *Homo* from Africa', *Annual Review of Anthropology* 33 (2004), 271–96.

N. M. Ashton, J. Cook, S. G. Lewis and J. Rose, *High Lodge: Excavations by G. de G Sieveking, 1962–8, and J. Cook, 1988* (British Museum Press, London, 1992)

N. M. Ashton, S. G. Lewis and S. A. Parfitt, 'High Lodge 2002', *Proceedings of the Suffolk Institute of Archaeology and Natural History* 40 (part 3; forthcoming)

J. M. Bermúdez de Castro, E. Carbonell and J. L. Arsuaga (eds.), 'Gran Dolina Site: TD6 Aurora Stratum (Burgos, Spain)', *Journal of Human Evolution* 37 (1999; special issue), 309–700

J. M. Bermúdez de Castro, M. Martinón-Torres, E. Carbonell, S. Sarmiento, A. Rosas, J. van der Made and M. Lozano, 'The Atapuerca Sites and their Contribution to the Knowledge of Human Evolution in Europe', *Evolutionary Anthropology* 13 (2004), 24–41

R. Dennell and W. Roebroeks, 'An Asian Perspective on Early Human Dispersal from Africa', *Nature* 438 (2005), 1099–104.

L. Gabunia, S. C. Antón, D. Lordkipandze, A. Vekua, A. Justus and C. C. Swisher III, 'Dmanisi and Dispersal', *Evolutionary Anthropology* 10 (2001), 158–70.

C. Gamble, 'Time for Boxgrove Man', *Nature* 369 (1994), 275–6

R. Gore, 'Dawn of Humans: Expanding Worlds', *National Geographic* (May 1997), 84–109

R. Gore, Dawn of Humans: The First Europeans', *National Geographic* (July 1997), 96–113

J. R. Lee, J. Rose, J. O. Hamblin and B. S. P. Moorlock, 'Dating the Earliest Lowland Glaciation of Eastern England: A Pre-MIS 12 Early Middle Pleistocene Happisburgh Glaciation', *Quaternary Science Reviews* 23 (2004), 1551–66

J. R. Lee, J. Rose, I. Candy and R. W. Barendregt, 'Sea-level Changes, River Activity, Soil Development and Glaciation around the Western Margins of the Southern North Sea Basin during the Early and Early Middle Pleistocene: Evidence from Pakefield, Suffolk, UK', *Journal of Quaternary Science* 21 (2005) 155–79

S. Lewis, C. Whiteman and R. Preece (eds.), *Norfolk and Suffolk Field Guide* (Quaternary Research Association, London, 2000)

G. Manzi, 'Human Evolution at the Matuyama–Brunhes Boundary', *Evolutionary Anthropology* 13 (2004), 11–24

S. Parfitt, R. Barendregt, M. Breda, I. Candy, M. Collins, G. R. Coope, P. Durbidge, M. Field, J. Lee, A. Lister, R. Mutch, K. Penkman, R. Preece, J. Rose, C. Stringer, R. Symmons, J. Whittaker, J. Wymer and A. Stuart, 'The Earliest Record of Human Activity in Northern Europe', *Nature* 438 (2005), 1008–12

G. Philip Rightmire, D. Lordkipanidze and A. Vekua, 'Anatomical Descriptions, Comparative Studies and Evolutionary Significance of the Hominin Skulls from Dmanisi, Republic of Georgia', *Journal of Human Evolution* 50 (2006), 115–41

W. Roebroeks and T. van Kolfschoten, 'The Earliest Occupation of Europe: A Short Chronology', *Antiquity* 68 (1994), 489–503

W. Roebroeks, 'Hominid Behaviour and the Earliest Occupation of Europe: An Exploration', *Journal of Human Evolution* 41 (2001), 437–61

D. C. Schreve (ed.), *The Quaternary Mammals of Southern and Eastern England Field Guide* (Quaternary Research Association, London, 2004)

CHAPTER TWO
UNDERSTANDING ICE AGES

P. Andrews, J. Cook, A. Currant and C. Stringer (eds.), *Westbury Cave: The Natural History Museum Excavations 1976–1984* (Western Academic and Specialist Press, Bristol, 1999)

J. Hays, J. Imbrie and N. Shackleton, 'Variations in the Earth's Orbit: Pacemaker of the Ice Ages', *Science* 194 (1976), 1121–32

S. Lewis, C. Whiteman and R. Preece (eds.), *Norfolk and Suffolk Field Guide* (Quaternary Research Association, London, 2000)

R. Meyrick and D. Schreve (eds.), *Central Germany (Thuringia) Field Guide* (Quaternary Research Association, London, 2002)

J. Murton, C. Whiteman, M. Bates, D. Bridgland, A. Long, M. Roberts and M. Waller (eds.), *Kent and Sussex Field Guide* (Quaternary Research Association, London, 1998)

S. Parfitt, 'A Butchered Bone from Norfolk: Evidence for Very Early Human Presence in Britain', *Archaeology International* 8 (2005), 15–18

M. Roberts and S. Parfitt (eds.) *Boxgrove: A Middle Pleistocene Hominid Site at Eartham Quarry, Boxgrove, West Sussex* (English Heritage Archaeological Report 17, London, 1999)

D. C. Schreve (ed.), *The Quaternary Mammals of Southern and Eastern England Field Guμide* (Quaternary Research Association, London, 2004)

C. Stringer, E. Trinkaus, M. Roberts, S. Parfitt and R. Macphail, 'The Middle Pleistocene Human Tibia from Boxgrove', *Journal of Human Evolution* 34 (1998), 509–47.

J. Wymer, *The Lower Palaeolithic Occupation of Britain* (Trust for Wessex Archaeology, Salisbury, 1999)

CHAPTER THREE THE GREAT INTERGLACIAL

J. L. Arsuaga, J. M. Bermúdez de Castro and E. Carbonell (eds.), 'The Sima de los Huesos Hominid Site', *Journal of Human Evolution* 33 (1997), 105–421

N. M. Ashton, J. McNabb, B. Irving, S. G. Lewis and S. Parfitt, 'Contemporaneity of Clactonian and Acheulian Flint Industries at Barnham, Suffolk', *Antiquity* 68 (1994), 585–9

N. M. Ashton, S. G. Lewis and S. A. Parfitt, 'Hoxne 2003', *Proceedings of the Suffolk Institute of Archaeology and Natural History* (forthcoming)

J. M. Bermúdez de Castro, M. Martinón-Torres, E. Carbonell, S. Sarmiento, A. Rosas, J. van der Made and M. Lozano, 'The Atapuerca Sites and their Contribution to the Knowledge of Human Evolution in Europe', *Evolutionary Anthropology* 13 (2004), 24–41

D. R. Bridgland, D. C. Schreve, D. H. Keen, R. Meyrick and R. Westaway, 'Biostratigraphical Correlation between the Late Quaternary Sequence of the Thames and Key Fluvial Localities in Central Germany', *Proceedings of the Geologists' Association* 115 (2004), 125–40

K. Duff (ed.), *The Story of Swanscombe Man* (Kent County Council and Nature Conservancy Council, Canterbury, 1985)

C. Gamble and M. Porr (eds.), *The Hominid Individual in Context: Archaeological Investigations of Lower and Middle Palaeolithic Landscapes, Locales and Artefacts* (Routledge, London, 2005)

S. Lewis, C. Whiteman and R. Preece (eds.), *Norfolk and Suffolk Field Guide* (Quaternary Research Association, London, 2000)

R. Meyrick and D. Schreve (eds.), *Central Germany (Thuringia) Field Guide* (Quaternary Research Association, London, 2005)

J. Murton, C. Whiteman, M. Bates, D. Bridgland, A. Long, M. Roberts and M. Waller (eds.), *Kent and Sussex Field Guide* (Quaternary Research Association, London, 1998)

M. Pitts, 'Who Ate the Elephant?', *British Archaeology* 80 (2005), 28–9

D. C. Schreve (ed.), *The Quaternary Mammals of Southern and Eastern England Field Guide* (Quaternary Research Association, London, 2004)

M. J. White and S. J. Plunkett, *Miss Layard Excavates: A Palaeolithic Site at Foxhall Road, Ipswich, 1903–1905* (Western Academic and Specialist Press, Bristol, 2004)

CHAPTER FOUR DESERTED BRITAIN

S. Aldhouse-Green (ed.), *Pontnewydd and the Elwy Valley Caves* (University of Wales Press and National Museums and Galleries of Wales, Cardiff, forthcoming)

N. Ashton and S. Lewis, 'Deserted Britain: Declining Populations in the British Late Middle Pleistocene', *Antiquity* 76 (2002), 388–96

N. Ashton, 'Absence of Humans in Britain during the Last Interglacial (Oxygen Isotope Stage 5e)', in A. Tuffreau and W. Roebroeks (eds.), *Le Dernier Interglaciaire et les Occupations Humaines du Paléolithique Moyen* (Publications du CERP, Lille, 2002)

N. Ashton, R. Jacobi and M. White, 'The Dating of Levallois Sites in West London', *Quaternary Newsletter* 99 (2003), 25–32

D. R. Bridgland, D. C. Schreve, D. H. Keen, R. Meyrick and R. Westaway, 'Biostratigraphical Correlation between the Late Quaternary Sequence of the Thames and Key Fluvial Localities in Central Germany', *Proceedings of the Geologists' Association*, 115 (2004), 125–40

P. Callow and J. Cornford (eds.), *La Cotte de St. Brelade 1961–1978* (Geo Books, Norwich, 1986)

A. P. Currant and R. M. Jacobi, 'A Formal Mammalian Biostratigraphy for the Late Pleistocene of Britain', *Quaternary Science Reviews* 20 (2001), 1707–16

P. L. Gibbard and J. P. Lautridou, 'The Quaternary History of the English Channel: An Introduction', *Journal of Quaternary Science* 18 (2003), 195–9

H. S. Green (ed.), *Pontnewydd Cave: A Lower Palaeolithic Hominid Site in Wales* (National Museum of Wales, Cardiff, 1984)

R. Preece (ed.), *Island Britain: A Quaternary Perspective* (The Geological Society, London, 1995)

D. C. Schreve, 'Differentiation of the British Late Middle Pleistocene Interglacials: The Evidence from Mammalian Biostratigraphy', *Quaternary Science Reviews* 20 (2001), 1693–705

D. C. Schreve (ed.), *The Quaternary Mammals of Southern and Eastern England Field Guide* (Quaternary Research Association, London, 2004)

D. C. Schreve, D. R. Bridgland, P. Allen, J. J. Blackford, C. P. Gleed-Owen, H. I. Griffiths, D. H. Keen and M. J. White, 'Sedimentology, Palaeontology and Archaeology of Late Middle Pleistocene River Thames Deposits at Purfleet, Essex, UK', *Quaternary Science Reviews* 21 (2002), 1423–64

D. C. Schreve, P. Harding, M. J. White, D. R. Bridgland, P. Allen, F. Clayton and D. H. Keen, 'A Levallois Knapping Site at West Thurrock, Lower Thames, UK: Its Quaternary Context, Environment and Age', *Proceedings of the Prehistoric Society*, forthcoming

A. Tuffreau and W. Roebroeks (eds.), *Bilans sédimentaires et occupations humaines durant le dernier interglaciare en Europe et en Proche Orient* (CERP, Lille, 2002)

M. J. White and D. C. Schreve, 'Island Britain – Peninsular Britain: Palaeogeography, Colonisation and the Earlier Palaeolithic Settlement of the British Isles', *Proceedings of the Prehistoric Society* 66 (2000), 1–28

M. White and N. Ashton, 'Lower Palaeolithic Core Technology and the Origins of the Levallois Method in NW Europe', *Current Anthropology* 44 (2003), 598–609

CHAPTER FIVE
NEANDERTHALS AND US

(various authors) 'Neanderthals Meet Modern Humans', *Athena Review* 2(4) (2001), 1–64

S. Aldhouse-Green (ed.), *Paviland Cave and the 'Red Lady': A Definitive Report* (Western Academic and Specialist Press, Bristol, 2000)

S. Aldhouse-Green, 'Great Sites: Paviland Cave', *British Archaeology* 61 (2001), 20–24

T. van Andel and W. Davies (eds.), *Neanderthals and Modern Humans in the European Landscape during the Last Glaciation* (McDonald Institute Monographs, Cambridge, 2003)

W. Boismier, 'Lynford Quarry: A Neanderthal Butchery Site', *Current Archaeology* 182 (2002), 53–8

W. Boismier, D. C. Schreve, M. J. White, D. A. Robertson, A. J. Stuart, S. Etienne, J. Andrews, G. R. Coope, M. Field, F. M. L. Greeh, D. H. Ken, S. G. Lewis, C. A. French, E. Rhodes, J.-L. Schwenninger, K. Tovey and S. O'Connor, 'A Middle Palaeolithic Site at Lynford Quarry, Mundford, Norfolk: Interim Statement', *Proceedings of the Prehistoric Society* 69 (2003), 315–24

R. Gore, 'Dawn of Humans: Neanderthals', *National Geographic* (January 1996), 2–35

R. Gore, 'Dawn of Humans: People Like Us', *National Geographic* (July 2000), 90–117

K. Harvati, S. R. Frost and K. P. McNulty, 'Neanderthal Taxonomy Reconsidered: Implications of 3D Primate Models of Intra- and Interspecific Differences', *Proceedings of the National Academy of Sciences of the USA* 101 (2004), 1147–52

K. Harvati and T. Harrison (eds.), *Neanderthals Revisited: New Approaches and Perspectives* (Springer, New York, forthcoming)

P. Mellars, 'Neanderthals and the Modern Human Colonization of Europe', *Nature* 432 (2004), 461–5

P. Mellars, C. Stringer, O. Bar-Yosef and K. Boyle (eds.), *Rethinking the Human Revolution* (McDonald Institute Monographs, Cambridge, forthcoming)

M. P. Richards, P. B. Pettitt, M. C. Stiner and E. Trinkaus, 'Stable Isotope Evidence for Increasing Dietary Breadth in the European Mid-Upper Paleolithic', *Proceedings of the National Academy of Sciences of the USA* 98 (2001), 6528–32

D. C. Schreve (ed.), *The Quaternary Mammals of Southern and Eastern England Field Guide* (Quaternary Research Association, London, 2004)

C. Stringer, 'The Neanderthal–*H. sapiens* interface in Eurasia', in K. Harvati and T. Harrison (eds.), *Neanderthals Revisited: New Approaches and Perspectives* (Springer, New York, forthcoming)

C. Stringer and W. Davies, 'Those Elusive Neanderthals', *Nature* 410 (2001), 791–2

C. Stringer, H. Pälike, T. van Andel, B. Huntley, P. Valdes and J. Allen, 'Climatic Stress and the Extinction of the Neanderthals', in T. van Andel and W. Davies (eds.), *Neanderthals and Modern Humans in the European Landscape during the Last Glaciation* (McDonald Institute Monographs: Cambridge, 2003)

C. B. Stringer, 'Out of Africa – A Personal History', in M. Nitecki (ed.), *Origins of Anatomically Modern Humans* (Plenum Press, New York, 1994)

M. J. White, 'The Stone Tool Assemblage from Lynford Quarry Mundford, and its Implications for Neanderthal Behaviour in Late Middle Palaeolithic Britain', in W. Boismier (ed.), *A Middle Palaeolithic Site at Lynford Quarry, Mundford, Norfolk* (ERAUL, Liège, forthcoming)

M. J. White and R. M. Jacobi, 'Two Sides to Every Story: Bout Coupé Handaxes Revisited', *Oxford Journal of Archaeology* 21(2) (2002), 109–33

CHAPTER SIX WHAT THEY GORGED IN CHEDDAR

P. Andrews and Y. Fernández-Jalvo, 'Cannibalism in Britain: Taphonomy of the Creswellian (Pleistocene) Faunal and Human Remains from Gough's Cave (Somerset, England)', *Bulletin of the Natural History Museum Geology Series* 58(supp) (2003), 59–81

L. Barham, P. Priestley and A. Targett, *In Search of Cheddar Man* (Tempus, Stroud, 1999)

R. N. E. Barton, R. M. Jacobi, D. Stapert and M. Street, 'The Late Glacial Reoccupation of the British Isles and the Creswellian', *Journal of Quaternary Science* 18 (2003), 631–43

A. Currant, R. Jacobi and C. B. Stringer, 'Excavations at Gough's Cave, Somerset 1986–7', *Antiquity* 63 (1989), 131–6

L. Humphrey and C. Stringer, 'The Human Cranial Remains from Gough's Cave (Somerset, England)', *Bulletin of the Natural History Museum Geology Series* 58 (2002), 153–68

R. M. Jacobi, 'The Late Upper Palaeolithic Lithic Collection from Gough's Cave, Cheddar, Somerset, and Human Use of the Cave', *Proceedings of the Prehistoric Society* 70 (2005), 1–92

R. M. Jacobi, 'The Stone Age Archaeology of Church Hole, Creswell Crags, Nottinghamshire', in P. B. Pettitt, P. Bahn and S. Ripoll (eds.), *Palaeolithic Cave Art at Creswell Crags in European Context* (Oxford University Press, Oxford, forthcoming)

P. B. Pettitt, P. Bahn and S. Ripoll (eds.), *Palaeolithic Cave Art at Creswell Crags in European Context* (Oxford University Press, Oxford, forthcoming)

A. W. G. Pike, M. Gilmore, P. B. Pettitt, R. Jacobi, S. Ripoll, P. Bahn and F. Muñoz, 'Independent U-Series Verification of the Pleistocene Antiquity of the Palaeolithic Cave Art at Creswell Crags, UK', *Journal of Archaeological Science* 32 (2005), 1649–55

M. Richards, R. Hedges, R. Jacobi, A. Currant and C. Stringer, 'Gough's Cave and Sun Hole Cave', *Journal of Archaeological Science* 27 (2000), 1–3

M. Richards, R. Jacobi, J. Cook, P. B. Pettitt and C. Stringer, 'Isotope Evidence for the Intensive Use of Marine Foods by Late Upper Palaeolithic Humans', *Journal of Human Evolution* 49 (2005), 390–94

S. Ripoll, F. Muñoz, P. Bahn and P. B. Pettitt, 'Palaeolithic Cave Engravings at Creswell Crags, England', *Proceedings of the Prehistoric Society* 70 (2004), 93–105.

D. C. Schreve (ed.), *The Quaternary Mammals of Southern and Eastern England Field Guide* (Quaternary Research Association, London, 2004)

C. Stringer, 'The Gough's Cave Human Fossils: An Introduction', *Bulletin of The Natural History Museum Geology Series* 56 (2000), 135–9

B. Sykes, *The Seven Daughters of Eve* (Bantam, London, 2001)

E. Trinkaus, L. Humphrey, C. Stringer, S. Churchill and R. Tague, 'Gough's Cave 1 (Somerset, England): An Assessment of the Sex and Age at Death', *Bulletin of the Natural History Museum Geology Series* 58(supp) (2003), 45–50

CHAPTER SEVEN OUR CHALLENGING CLIMATES

(various authors) 'The Heat is On', *National Geographic* (September 2004), 2–75

(various authors) 'Heat: How Global Warming is Changing Our World', *Guardian* (supplement), 30 June 2005

(various authors) 'Turning the Tide', *Observer/Carbon Trust* (supplement), 26 June 2005

R. Alley, *The Two-Mile Time Machine: Ice Cores, Abrupt Climate Change and Our Future* (Princeton University Press, Princeton, NJ, 2002)

J. Diamond, *Collapse: How Societies Choose to Fail or Succeed* (Allen Lane, London, 2005)

B. Fagan, *The Long Summer: How Climate Changed Civilization* (Granta Books, London, 2004)

T. Flannery, *The Weather Makers: The History and Future Impact of Climate Change* (Allen Lane, London, 2006)

R. A. Kerr, 'News Focus – Three Degrees of Consensus', *Science* 305 (2004), 932–4

J. Lovelock, *The Revenge of Gaia* (Allen Lane, London, 2006)

P. N. Pearson and M. R. Palmer, 'Atmospheric Carbon Dioxide Concentrations Over the Past 60 Million Years', *Nature* 406 (2000), 695–9

R. T. Pinker, B. Zhang and E. G. Dutton, 'Do Satellites Detect Trends in Surface Solar Radiation?', *Science* 308 (2005), 850–54

D. L. Royer *et al.*, 'Paleobotanical Evidence for Near Present-day Levels of Atmospheric CO_2 During Part of the Tertiary', *Science* 292 (2001), 2310–13

W. Ruddiman, 'How Did Humans First Alter Global Climate?', *Scientific American* Vol. 292, No. 3 (2005), 34–41

H. J. Schellnhuber, W. Cramer, N. Nakicenovic, T. Wigley and G. Yohe (eds.), *Avoiding Dangerous Climate Change* (Cambridge University Press, Cambridge, 2006)

M. Wild, H. Gilgen, A. Roesch, A. Ohmura, C. Long, E. Dutton, B. Forgan, A. Kallis, V. Russak and A. Tsvetkov, 'From Dimming to Brightening: Decadal Changes in Solar Radiation at Earth's Surface', *Science* 308 (2005), 847–850

K. Willis, K. Bennett and D. Walker (eds.), *The Evolutionary Legacy of the Ice Ages* (Royal Society, London, 2004)

INDEX

Figures in italics refer to illustrations

BRENTFORD LONDON
GTRLONDON

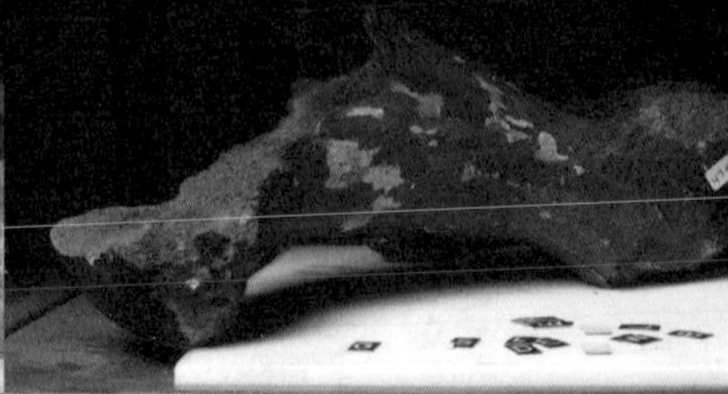
FELTHAM

PALL MALL

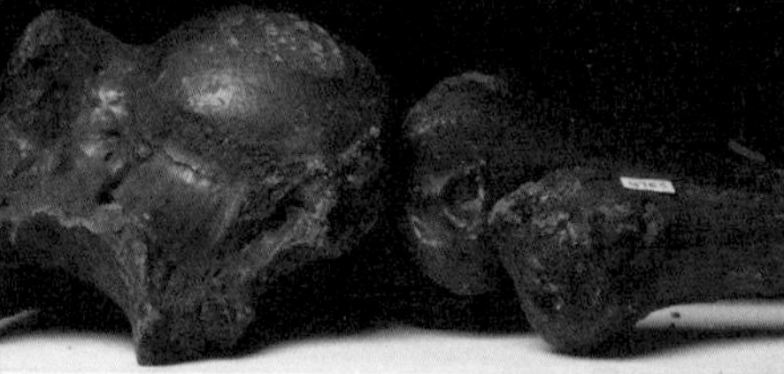
OLD KENT ROAD

SHEPPERTON

BATTERS

NEW ZEALAND HOUSE

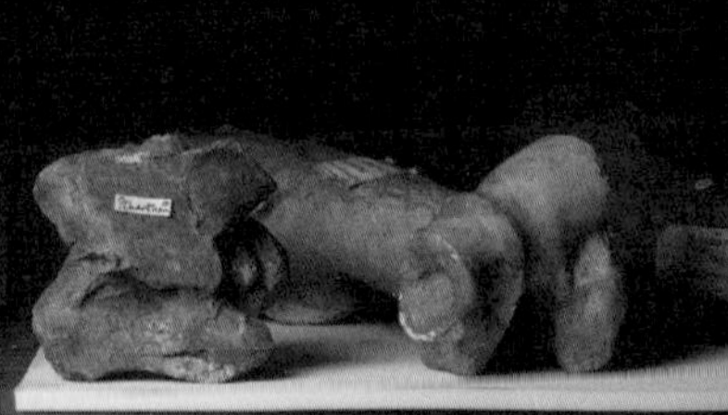
BUCKINGHAM PALACE ROAD
GREENHAMS
PI

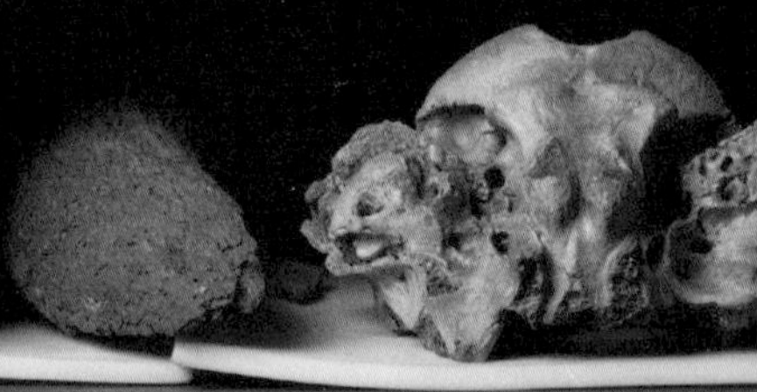
TRAFALGAR SQ
MITCHAM